Crocosmia and *Chasmanthe*

Royal Horticultural Society Plant Collector Guide

Crocosmia and *Chasmanthe*

Peter Goldblatt
John Manning
Gary Dunlop

Illustrations by
Auriol Batten

Timber Press
Portland · Cambridge

ROYAL HORTICULTURAL SOCIETY

The Royal Horticultural Society is the world's leading charity and membership organisation working to provide inspiration, advice and information for gardeners of all levels and to promote excellence in horticulture. Its activities include demonstration gardens, flower shows and events across the United Kingdom, research and advice, plant trials, and publications.

An interest in gardening is all you need to enjoy being a member of the RHS. For more information visit our Web site, www.rhs.org.uk, or call 0845-130-4646 in the United Kingdom.

Registered charity number 222879

Copyright © 2004 by Peter Goldblatt, John Manning and Gary Dunlop
All rights reserved.

All color paintings (reproduced at a scale of approximately ×0.85) and drawings by Auriol Batten unless otherwise indicated.

Published in association with the Royal Horticultural Society in 2004 by

Timber Press, Inc.
The Haseltine Building
133 S.W. Second Avenue, Suite 450
Portland, Oregon 97204-3527, U.S.A.

www.timberpress.com

Timber Press
2 Station Road
Swavesey
Cambridge CB4 5QJ, U.K.

Series design by Dick Malt; volume design by Susan Applegate
Printed through Colorcraft Ltd., Hong Kong

Library of Congress Cataloging-in-Publication Data

Goldblatt, Peter, 1943–
    Crocosmia and Chasmanthe / Peter Goldblatt, John Manning, Gary Dunlop.
        p. cm. — (Royal Horticultural Society plant collector guide)
    Includes bibliographical references.
    ISBN 0-88192-651-5 (hardcover)
    1. Crocosmia. 2. Chasmanthe. I. Manning, John. II. Dunlop, Gary. III. Title. IV. Series
    QK495.175G6199 2004
    584′.38—dc22   2004000119

A catalogue record for this book is also available from the British Library.

Contents

CONTENTS

Color plates follow page 80

Preface

Plants of increasing importance in horticulture, both for garden display and the cut-flower trade, little has been written about the genera *Chasmanthe* and *Crocosmia*. There are scattered articles in horticultural journals dealing with a handful of cultivars of *Crocosmia* and virtually none on *Chasmanthe*. Likewise, accounts of the classification of *Chasmanthe* and *Crocosmia* are few, published in the scientific literature, presented in specialized botanical terminology and thus not readily available to non-specialists. Moreover, until now the most recent complete account of *Crocosmia*, written by the South African botanist Miriam de Vos, was published in 1984. Her classification was repeated in a treatment of the genus for the *Flora of Southern Africa*, published in 1999, long after the manuscript was completed. These studies are now somewhat outdated, owing to the accumulation of new collections from the wild, a revised understanding of some species and the addition of the Madagascan *Crocosmia ambongensis* to the genus.

From a horticultural perspective, *Chasmanthe* and *Crocosmia* seem unnecessarily neglected, though there is an accumulating literature on *Crocosmia* and a growing appreciation of the plants' value as reliable garden subjects in diverse climates for summer and fall color. In season, cultivars of *Crocosmia* are also sold as cut flowers. *Chasmanthe* species, were they better known, would extend the season to spring and even late winter in some parts of the world. Yet there is no manual on *Crocosmia* nor any complete account of its history even though the genus was first introduced into cultivation in the mid-1840s and after 1879 was subject to intensive crossbreeding and selection.

This book aims to fill multiple needs, which include a scientific treatment of all the wild species, a history of their discovery and introduction to Western horticulture and a detailed narrative of the history of *Crocosmia* breeding and the economic forces that changed that direction. Available nowhere else, this book includes as full a list of *Crocosmia* cultivars as possible.

Crocosmia and Chasmanthe has from the start been a collaborative project. It was originally conceived by the Irish botanist Charles Nelson and one of us, John Manning, to continue a series of popular but scientifically accurate volumes on various horticulturally significant genera of the *Iris* family, combining botanical expertise with the work of talented botanical artists. Examples of such works include *The Moraeas of Southern Africa* by Peter Goldblatt with the artist Fay Anderson (Kirstenbosch National Botanical Garden, 1986), *The Genus Watsonia* by Goldblatt, using archival paintings by various artists and new paintings by Fay Anderson (Kirstenbosch National Botanical Garden, 1989), *The Woody Iridaceae* by Goldblatt with Anderson (Timber Press, 1993), *Gladiolus in Tropical Africa* by Goldblatt with illustrations by Manning (Timber Press, 1996) and *Gladiolus in Southern Africa* by Goldblatt and Manning, with paintings by Fay Anderson and Auriol Batten (Fernwood Press and Timber Press, 1998). I was originally asked to be the third author of this trio while the artists Auriol Batten in South Africa and Wendy Walsh in Ireland were invited to contribute paintings. Our initial plan to use plants already in cultivation in Great Britain or South Africa for illustration soon proved unworkable. Plants in cultivation are not always representative of the wild plants, and several species were simply not available at all.

Two *Chasmanthe* species grow wild around Cape Town and proved easy to collect in flower and fruit. A third was not known in the wild, and we used plants growing at Kirstenbosch for description and illustration.

For *Crocosmia*, the larger genus, obtaining wild plants in flower proved a more daunting undertaking. *Crocosmia aurea* was easily obtained because its wide range includes East London, South Africa, where Auriol Batten could collect samples easily. *Crocosmia fucata* is, however, rare and not in cultivation. We had to locate a wild population in the higher mountains of Namaqualand and then mail samples as quickly as possible to East London. The South African mail service proved invaluable; carefully packaged plants collected hundreds of miles from East London always arrived fresh and undamaged not more than 24 hours after being mailed and often no more than 48 hours after collecting in such diverse places as Namaqualand in Northern Cape province, God's Window Nature Reserve in Mpumalanga, Howick in KwaZulu-Natal and The Sentinel in Free State province.

Crocosmia masoniorum proved particularly difficult to find. Although it grows just 200 miles (320 km) from East London, it is nowhere common and is reported from just three sites in the mountainous Drakensberg of the Transkei. Finally, I tracked down a population at, I suspect, the locality where it was first discovered in 1896 and collected there again in 1911. My success in finding *C. masoniorum* owes much to the detailed description of the habitat provided by Marianne Mason (the species was named for her and her brother, who together gathered plants in 1911), which is preserved in correspondence at the herbarium at the Royal Botanic Gardens, Kew. Plants were in fruit in late summer (February) when I found them and took them immediately to East London. The following year John Manning followed in my tracks and in early January 2003 collected flowering specimens for Auriol Batten to paint.

Both Charles Nelson and Wendy Walsh found it impossible to maintain their commitment to the project and bowed out. Auriol immediately agreed to complete paintings not only of our wild-collected plants but also of the cultivars, but only with Wendy Walsh's blessing. That was readily obtained, and with the generous help of Gary Dunlop in providing corms of important cultivars, Auriol first grew and then painted the cultivars illustrated here. Without Charles's participation, we needed a collaborator who knew crocosmias well, and Gary was invited to contribute both his horticultural expertise and knowledge of the history of crocosmia breeding. His willingness to devote time and energy to our cause has resulted in a book that is scientifically and historically as accurate as possible, given the limitations that the sparse written record permits.

In *Crocosmia and Chasmanthe* we have compiled a volume with many features. It is a history of plant exploration and discovery of two genera and their introduction into cultivation. But it is also a thorough account of how the species and the cultivars should be grown. In addition, it contains a full botanical account of *Chasmanthe* and *Crocosmia,* a history of the breeding of *Crocosmia,* and a complete listing of the numerous cultivars raised over a period of some 125 years, since the first hybrids were made by Victor Lemoine in France in 1879.

PETER GOLDBLATT

Acknowledgments

We thank those who have graciously helped us in various ways, particularly Graham Duncan at Kirstenbosch National Botanical Garden, Cape Town; Olive Hilliard of the Royal Botanic Garden, Edinburgh; Peter Phillipson of the Missouri Botanical Garden, Saint Louis; Brenda Solomon of Johannesburg; Alan Tait, Mary Bursey, Lynne Johnston and Carl Vernon, all of East London; Mervyn Lötter of the Mpumalanga Department of Nature Conservation, who helped with plant material; Cameron McMaster of Stutterheim, South Africa; Ingrid Nänni of the National Botanical Institute, South Africa; Pieter Winter of the University of the North, Limpopo province, South Africa; Jasmine Beukes of the farm Modderfontein, where *Crocosmia fucata* grows; Roy Gereau and Mary Stiffler of the Missouri Botanical Garden; and Lendon Porter of Portland, Oregon. All assisted in various ways to obtain plants or provide advice or library requests, or companionship and assistance in the field. It has been a pleasure to work with all of them.

We particularly thank David Fenwick of Plymouth in the United Kingdom for readily sharing his ample information about *Crocosmia* cultivars and their history. Our editor at Timber Press generously offered helpful comments and advice throughout the preparation of this volume, which is much improved as a result.

Last, but not least, we must express our deep appreciation to Auriol Batten for her enthusiasm and fine work. As in past projects, it has been a delight to work with her.

The compilation of information about the *Crocosmia* cultivars was assisted by the librarian and staff of the Royal Horticultural Society's Lindley Library, and the Botany Library, Natural History Museum, London;

Margaret Andrew (research assistant) and subsequently Jill Hutchens, both of the National Council for the Conservation of Plants and Gardens of the United Kingdom, for researching and supplying copies of RHS awards and various nursery catalogues); Phillip Wood, retired director, Slieve Donard Nursery; Peter Borlase, original National Collection Holder of *Crocosmia,* now retired, Lanhydrock Garden, Cornwall; Jonathon and Daphne Shackleton; Helen Dillon; Jim Mahir, *Crocosmia* collector and breeder, Dublin, Ireland; Martyn Rix; James Compton; Jill Cowley, herbarium at the Royal Botanic Gardens, Kew, now retired; Craig Brough, enquiries librarian, Kew; Phillipa Browne; Chris Saunders; Alan Street; Lynda Windsor; Robin White; Wim Snoeijer, the Netherlands, for researching nursery catalogues at the Koninklijke Algemeene Vereeniging voor Bloembollencultuur; Peter Yeo, Cambridge University Botanic Garden, now retired; the late Molly Sanderson; Bill Lennon; Ronnie Cameron; and lastly but not least, Andrew Templeton, for translating the descriptions of the French cultivars.

A publication subsidy for *Crocosmia and Chasmanthe* was provided by the Stanley Smith Horticultural Trust, making possible the production of this volume at reasonable cost without sacrificing quality. We are grateful to this organization for their support. Also, we gratefully acknowledge support for fieldwork from grants 6704-00 and 7316-02 from the U.S. National Geographic Society.

Early Exploration and Discovery

Chasmanthe

Native to the Cape region, the extreme southwestern tip of South Africa, chasmanthes have a longer known history than crocosmias, which occur widely across sub-Saharan Africa. This is because botanical exploration of the African continent south of the Sahara began in the 16th century when Portuguese and then Dutch explorers pioneered sea travel around the southern tip of Africa to the fabled Indies. The Dutch, who established a resting and provisioning station at the Cape in 1652, later to become Cape Town, were even in those days deeply curious about plants, and ship captains were encouraged to bring back curiosities that might be grown for interest and, of course, profit in Holland. Bulbous plants were ideal subjects for shipping back to Europe in sailing ships returning in the late southern summer, as they followed the southeastern trade winds from the Indian Ocean to the Atlantic Ocean. As early as the beginning of the 15th century, Cape bulbs found their way to Holland, Germany and England, and were flowered in greenhouses, engendering intense interest, both horticultural and scientific, a phenomenon described in detail by Manning et al. (2002).

Among the earliest of these bulbs were the species of what we now call *Chasmanthe*. The first record of a *Chasmanthe* species, which we now know as *C. floribunda*, appeared in a 1635 volume by the French physician and scientist Jacques-Philippe Cornut (1606?–1651), *Canadensium Plantarum, Aliarúmque Nondum Editarum Historia* (history of Canadian plants and of others not yet published), in which there is a fair woodcut, unmistakably representing this plant, under the delightful name *Gladiolus aethiopicus flore coccineo*. This multiword name is called a polynomial and was the way plants were named until 1753 when the Swedish botanist, zoologist, physician and originator of the current system of naming organisms, Carl Linnaeus, established the binomial system of nomenclature for plants and animals. Cornut's illustration was based on plants grown in Paris, which flowered there according to Cornut as early as 1633. This is quite remarkable when one realizes that France was then ruled by Louis XIII, Britain by Charles I, and the Cape route around the southern tip of African to the Indies had only been discovered by Bartolomeu Dias and Vasco da Gama at the end of the 15th century, less than 150 years earlier. North America had barely begun to be explored, and Massachusetts and Virginia were the only two colonies established there.

Cornut's illustration was copied by the British botanist and first professor of botany at Oxford, Robert Morison (1620–1683), in his 1680 work, *Plantarum Historiae Universalis Oxoniensis* (Oxford universal history of plants). A second illustration of a *Chasmanthe* species was published in a 1692 work by the English botanist Leonard Plukenet, *Phytogeographia* (plant geography).

Linnaeus, who knew *Chasmanthe* from these early illustrations and from specimens preserved in herbarium collections in Holland and England, formally named the plant *Antholyza aethiopica* in 1759, thus preserving part of Cornut's polynomial for the species. Inappropriate as the specific epithet now seems to us, until the 18th century all of Africa south of the Sahara was often referred to as Aethiopia. Only relatively recently was the name restricted to the northeastern African country now called Ethiopia. Unfortunately, Linnaeus mistakenly associated Cornut's illustration (and the inferior copies in later works), which is *C. floribunda*, with a second common Cape *Chasmanthe*, the autumn- and early-winter-flowering *C. aethiopica*. In fact, these two superficially similar plants were frequently confused with one another for the next two centuries, and the two names were inconsistently applied in the literature. Linnaeus believed that all the red-flowered southern African plants of the *Iris* family (plants with six tepals, three stamens and sword-shaped flat leaves) that had an elongate perianth tube, narrow at the base and wider above, and a prominent dorsal (upper) tepal belonged in one genus, which he established as *Antholyza*—hence his choice of name, *A. aethiopica*, for what is now *Chasmanthe aethiopica*.

Cornut's *Chasmanthe*, which is *C. floribunda*, remained in cultivation in France and England, and it was only in 1812 that the gardener-cum-botanist Richard A. Salisbury published his conclusion that there were two species in gardens in Great Britain then being called *Antholyza aethiopica*. We assume Salisbury had no way to effectively determine which of the two was the true *A. aethiopica*, and instead of making what we would today see as an arbitrary decision, he named one of the two *A. floribunda* and the other *A. vittigera*. No sooner had these names been published than the French botanist and plant illustrator Pierre Joseph Redouté published a fine painting of the tall species, *A. floribunda*, in 1813 under the name *A. prealta* in his monumental illustrated work, *Les Liliacées*. Most likely Redouté's account was already in press when Salisbury's name for the species was published, and we assume Redouté and Salisbury independently reached the conclusion that two different species were until then called *A. aethiopica*.

A third species of *Chasmanthe* only came into cultivation in Europe some time after 1800. Writing of this plant in 1828, the English botanist John Lindley said that it had long been grown in European gardens. Rare in the wild, just how this attractive species reached Europe is unrecorded, but herbarium specimens in the herbarium of the natural history museum in Vienna, an important center for the study of African flora, bear the date 1811. Using Linneaus's name *Antholyza*, Lindley called the plant *A. aethiopica* var. *minor* (*minor* meaning small); at that time the name *A. aethiopica* was still often applied the tall-stemmed and large-flowered *A. floribunda*. Plants grown in Boccadifalco near Palermo in Sicily, where there was a long-established botanic garden, were formally named *A. bicolor* in 1832 by the botanist Guglielmo Gasparrini in an account of some plants cultivated there. The species was also included in 1845 in a catalogue of the plants cultivated at the Royal Botanic Garden in Naples (*Catalogo delle Piante Che Si Coltivano nel R. Orto Botanico di Napoli*) by Michele Tenore. Although the name *A. bicolor* and the original description of the plant are often credited to Gasparrini's colleague Tenore, the botanical analysis of the species appears under Gasparrini's name in both the 1832 and 1845 accounts of *A. bicolor*. Sole authorship of the name *A. bicolor* must therefore correctly be credited to Gasparrini.

Antholyza, as we now understand the genus, consisted of disparate plants in the 18th and 19th centuries, similar mainly in their flowers, which are adapted to pollination by sunbirds. In current terms we would say *Antholyza* included a range of species of different lineages, independently adapted for bird pollination. Gradually, as their respective relationships became better understood, the species of *Antholyza* were assigned to genera that grouped the closest relatives together, usually those with less elaborate flowers. Thus some *Antholyza* species were removed to *Babiana, Watsonia* or *Gladiolus,* the latter including the type species of the genus *Antholyza, A. cunonia,* making *Antholyza* a nomenclatural synonym of *Gladiolus.* This was not widely appreciated at the time, and many botanists, including the influential English botanist and bulbous-plant specialist John Gilbert Baker, continued to recognize *Antholyza* as a distinct genus. By 1932, however, the genus was widely accepted as untenable in Linnaeus's sense, and a second English botanist, Nicholas Edward Brown, described *Chasmanthe* for three species of the erstwhile *Antholyza.*

Nomenclatural confusion did not end there. In 1941 the South African botanist Edwin Percy Phillips transferred all the *Chasmanthe* species to *Petamenes,* a genus in which he included not only *Chasmanthe* but several *Gladiolus* species and what is now *Crocosmia fucata.* Phillips's reason for this action was that he thought *Petamenes* was an earlier generic name for the species that he included in the genus. The type species of *Petamenes, P. abbreviatus* Andrews, is now included in *Gladiolus* (Goldblatt and Manning 1998), making *Petamenes* a synonym of the Linnaean genus *Gladiolus. Chasmanthe* has thus once again assumed the definition that N. E. Brown gave it. Today, *Chasmanthe* is recognized as allied to *Babiana, Ixia, Sparaxis* and *Tritonia,* but exactly how these genera are interrelated remains to be resolved. As we discuss in more detail in a later chapter, *Chasmanthe* appears on the basis of DNA evidence to be most closely related to *Babiana,* something no one would have predicted from morphology alone.

Lack of understanding of species and their patterns of variation led N. E. Brown to add several additional species to *Chasmanthe,* but all were either minor or local variants of *C. aethiopica* (for example, *C. peglerae* and *C. vittigera*) or actually proved on more careful examination to belong to other genera, including *Crocosmia, Gladiolus* and *Tritoniopsis.*

Crocosmia

The botanical history of *Crocosmia* offers less romance than that of *Chasmanthe* but a greater measure of taxonomic mystery. The first species of what we now regard as the genus *Crocosmia* to be discovered was the rare Namaqualand species *C. fucata*. Namaqualand, the semiarid country lying to the north of the Cape region, along the western coast and adjacent interior of South Africa, has a low rainfall, most falling in the winter (June and July), and is an unlikely provenance for a plant belonging to a genus otherwise of evergreen forest or well-watered grassland habitats in areas of relatively high summer rainfall. The Namaqualand plant was first referred to the genus *Tritonia* by the British bulb expert William Herbert in 1838 and, as Herbert relates, was evidently brought to England many years before then. Herbert records that he had grown plants in his garden in Yorkshire for some 25 years, initially under glass, where corms multiplied but did not flower. Corms left outdoors showed the plants to be hardy, but it was only after the application of manure in the autumn of 1836 that plants flowered in 1837. A flowering branch was immediately painted in watercolor, a medium much used at that time to record the appearance of a species. A formal description together with a reproduction of the painting was published the following year in the periodical *Edwards's Botanical Register*. We remain in the dark about how Herbert first obtained his plants, which must have reached Great Britain no later than 1812.

The first documented record of *Crocosmia fucata* in the wild has now been traced to the German plant collector and explorer Carl Ludwig Zeyher, who, in collaboration with the Dane Christian Friedrich Ecklon, made a living gathering plant specimens in southern Africa and selling them to European scientific institutions. Zeyher visited the high-elevation Kamiesberg of Namaqualand in 1829 (Gunn and Codd 1981), where he collected specimens of *C. fucata*, but this was clearly too late a date for Zeyher to have been the source of the plants grown by William Herbert. We speculate that *C. fucata* may have been collected by James Niven, the Scottish gardener and plant collector sent to the Cape by the wealthy English magnate George Hibbert in 1798. On a second visit to the Cape, Niven collected plants for Empress Josephine Buonaparte of France and the Lee and Kennedy Nursery in London until 1812, when he returned to Great Britain. Niven collected plants, including some bulbous species, in Namaqualand in 1799, but whether as bulbs, corms or seeds in unknown. Some of Niven's collections flowered years later and were only then described. *Moraea longiflora*, believed to be one of Niven's discoveries of 1799, flowered in London in 1804, when it was given by Hibbert to John Ker-Gawler to describe. *Crocosmia fucata* may have also been collected by Niven. Another possibility is that *C. fucata* was collected by the collector for the Royal Botanic Gardens, Kew—Francis Masson—who is known to have sent plants back to Britain from Namaqualand in the early 1790s. Zeyher's specimen of *C. fucata,* evidently collected in 1829 (though this is not indicated on the information accompanying his specimens), was from the Kamiesberg near 'Nieuw Kerksfonteyn' at 1200–1500 feet (370–460 m), thus at relatively low elevation, on the farm now called Niekerksfontein.

The native geographic range of *Crocosmia fucata* was finally confirmed only in 1910 when the expedition led by the botanist and first director of the National Botanical Garden at Kirstenbosch, Henry Herald Welch Pearson (1870–1916), found plants on the upper slopes of Sneeukop, second-highest peak in the Kamiesberg, which reaches approximately 5000 feet (1500 m). J. G. Baker transferred *Tritonia fucata* to *Antholyza* in 1877, and in 1932 N. E. Brown removed it to *Chasmanthe*. Such was the misunderstanding of the evolutionary relationships of this plant that it only found its present position in *Crocosmia* in 1984. This was after Miriam P. de Vos, who revised both *Crocosmia* (de Vos 1984) and then *Chasmanthe* (de Vos 1985), re-collected plants on the farm Modderfontein, which lies on the eastern slopes of Sneeukop. In November 1995 we (P.G. and J.M.) found large clumps of *C. fucata* growing in quantity along a seasonal stream, already dry that late in the season. De Vos decided that *C. fucata* belongs in *Crocosmia* because it matches that genus in its divaricately branched flowering stem, flexuose arching spike, stamens equal in length, apically forked style branches, capsules shorter than wide, and basic chromosome number $x = 11$. Species of *Chasmanthe* do not have the stem branched in this manner, the median filament is longer than the other two, the style branches are not forked apically, the capsules are ovoid, and the basic chromosome number $x = 10$.

Botanical exploration of Africa proceeded gradually from the Dutch colony at the Cape into the interior of the continent in the late 18th century, but it was only after the Cape became a British possession in the early 19th century that exploration accelerated. Thus it was only after 1820 that travelers reached eastern South Africa and then tropical Africa. The widespread *Crocosmia aurea* was evidently first collected by the German pharmacist and traveler-naturalist Johann Franz Drège, who made an epic journey from the barely settled eastern part of the Cape Colony, now the Eastern Cape province, in February 1832. Drège's route took him through the trackless and dangerous Transkei, inhabited by the restless and at time war-like Xhosa people, to Port Natal, now Durban, some 400 miles (640 km) to the east. His collections, including *Drège 4551* from near Umtata, were distributed somewhat later and thus were not immediately described. They do, however, document his discoveries, many of which were the first records of common African plants. The German botanist and explorer Wilhelm Peters collected *C. aurea* at Boror in Mozambique during his travels there in 1842–1848. Peters's collections, which were sent to Berlin, were written up later by colleagues, and its was only in the 1860s that the results of his important expedition were published. The collection of *C. aurea* was so identified (though with the genus spelled 'Crocosma') in the published work, but a manuscript name, 'Crocanthus mossambicensis Klotzsch', listed in the text, was associated with Peters's collection. This suggests to us that the German botanist Johann Friedrich Klotzsch (1805–1860) had reached the conclusion that the plant represented a new genus, possibly before the first species of what would become *Crocosmia* was described in 1847.

The collection that was to become the type species of *Crocosmia,* now *C. aurea,* was made in the 1840s, well after Drège had made his gathering in 1832. It was initially believed to have come from George in the southern Cape, 250 miles (400 km) east of Cape Town, where the plant was painted by the flower artist Jean Villet, who lived near this isolated settlement. The painting and a preserved specimen were sent to the Royal Botanic Gardens, Kew, where they were received in 1847. Miriam de Vos, who revised *Crocosmia* in 1984, thought that Villet may have made the original collection, but that seems unlikely though plants may have been grown at George, leading to the misunderstanding. We now believe that

'George' could have meant almost anywhere along the southern Cape coast from the present town of that name to Algoa Bay, some 180 miles (290 km) to the east. *Crocosmia aurea* is not known as a wild plant anywhere nearer to Cape Town than the vicinity of Grahamstown and Uitenhage, north of Algoa Bay, and we assume that this is where the original plants were found. The collection became the basis for *Tritonia aurea,* named in 1847 by the director of the Royal Botanic Gardens, Kew: Joseph Dalton Hooker. He used a manuscript name on the herbarium sheet at the Kew herbarium written by the Colonial Botanist at the Cape, the German-born Karl Wilhelm Ludwig Pappe (1803–1862). Pappe is sometimes credited with having made the type collection himself, and he evidently suggested the specific epithet, *aurea.* Evidence assembled by the compilers of a volume that details the botanical exploration of southern Africa (Gunn and Codd 1981), however, indicates that Pappe traveled to the Knysna area (not far from George) and to the Eastern Cape only after 1852. The collector of the type collection of *C. aurea* thus remains uncertain. Villet is however, responsible for the initial introduction of *C. aurea* into cultivation, and Hooker reported that the species was first cultivated by the James Backhouse Nursery at York in Great Britain.

The establishment of the Natal Colony, annexed to the British Empire in 1843, brought the plants of eastern South Africa to the world's attention. Many plants from the so-called Garden Colony were introduced into Europe from this botanically rich part of Africa. Most important of these was undoubtedly *Gladiolus dalenii,* the major contributor to the genetic stock of the cultivated *Gladiolus,* now a hugely important horticultural crop for the cut-flower and garden trade. Natal was also to yield additional stock of *Crocosmia aurea,* including the taller, more robust kind that was later much used in breeding *Crocosmia* cultivars.

The French physician and botanist Jules Émile Planchon described the genus *Crocosmia* in 1851, recognizing the single species *C. aurea,* which he transferred from *Tritonia.* The genus was eventually accepted by J. G. Baker in 1892 and 1896 and by the German plant specialist Friedrich Wilhelm Klatt (1894), both of whom spelled the name '*Crocosma*'. The original spelling of *Crocosmia* was queried by Rev. Charles Wolley-Dod (1882), who cited examples such as *Coprosma* (Rubiaceae) and *Diosma* (Rutaceae) where *-osma* (scented) is favored. While this argument seems grammatically sound, the original spelling *Crocosmia* is, nevertheless, currently accepted.

The source of the next *Crocosmia* to be discovered, *C. paniculata,* first of the pleated-leaved species of the genus to be described, remains a puzzle. Specimens said to be from 'Natal and Zululand', but without precise locality, were collected by William Tyrer Gerrard (d. 1866), together with Mark Johnston McKen (1823–1872), according to information on the sheet of the collection at the KwaZulu-Natal Herbarium, Durban. These two early Natal botanists collected *C. paniculata* between 1861 and 1865, when a preserved specimen reached the Royal Botanic Gardens, Kew. This is the type of what F. W. Klatt named *Antholyza paniculata* in 1867. That the form of the species that Gerrard and McKen collected is common in Mpumalanga province, adjacent to northern KwaZulu-Natal, but rare in KwaZulu-Natal itself makes the source of the type collection uncertain. *Antholyza paniculata* was transferred to the new genus *Curtonus* by N. E. Brown in 1932 when he dismantled *Antholyza* in his attempt to create a more natural classification of the disparate group of species then included in that genus. The affinities of *Curtonus* seemed clearly to be with *Crocosmia,* and on the basis of similar morphology and chromosome number, Goldblatt (1971) transferred *Curtonus paniculatus* to *Crocosmia.*

Mystery also surrounds the discovery of a third eastern southern African *Crocosmia, C. pottsii,* which was found in Natal in the 1870s. Named *Montbretia pottsii* in 1877 by J. G. Baker, then one of the world's experts on bulbous plants, especially African Iridaceae, *C. pottsii* was based on plants cultivated at the Royal Botanic Garden, Edinburgh. The gardener there, James McNab (1810–1878) had already compiled a list (unpublished) of outdoor plants cultivated in Edinburgh in which he listed the plant with the provisional name '*Gladiolus pottsii*'. According to Max Leichtlin, an influential horticulturist who maintained an important garden at Baden-Baden, Germany (Beissner 1892), the species was obtained from George Honington Potts (1830–1907; Desmond 1994) of Fettes Mount, Lasswade, near Edinburgh, who introduced it from South Africa. *Montbretia pottsii,* also called *Tritonia pottsii,* quickly found its way to the important nursery of Victor Lemoine in Nancy, France, where many plant cultivars were produced. Lemoine successfully bred early garden *Gladiolus,* including what became known as *G. ×lemoinei,* the progeny of a cross between early garden *Gladiolus* selections and the southern African *G. papilio* (see Goldblatt 1996).

Lemoine began to breed what we now call crocosmias as soon as his stock of *Montbretia* (*Crocosmia*) *pottsii* flowered, probably in 1879. Evidently he also grew *C. aurea* at the time, and he made the important interspecific hybrid between *C. aurea* and *C. pottsii*, which first flowered in August 1881. It is notable that Lemoine was growing a Natal plant, formally named only in 1877, but was already aware that it had been described in the genus *Montbretia*. Almost certainly he obtained the plant from Max Leichtlin, whose garden at Baden-Baden lay not far east of Lemoine's nursery at Nancy. Lemoine marketed his free-blooming and relatively large-flowered hybrid as '*Montbretia crocosmiaeflora*' (*Montbretia* with a *Crocosmia*-like flower), which rapidly became a popular garden subject and is now grown widely in warm temperate and subtropical parts of the world. Lemoine's 1882 catalogue shows the high value he placed on this plant. While corms of *C. pottsii* cost 1 franc each, a corm of '*M. crocosmiaeflora*' cost 6 francs; in 1882 that would have bought a grand meal at a fine restaurant.

Difficult to eliminate and evidently disease free, the vigorous *Crocosmia* ×*crocosmiiflora* (today written with a multiplication sign × before the species name to denote a hybrid) persists in abandoned gardens and spreads gradually into undisturbed woodland and forest margins. It is sometimes so persistent that it is treated as a weed in parts of Australia, the United States and New Zealand. This vigorous hybrid is now naturalized across central America, Madagascar, Indonesia, the Philippines, Hawaii and is even found as a weed many miles from any habitation in southern Africa. According to the taxonomy of the time, '*Montbretia crocosmiaeflora*' was an intergeneric hybrid, but this changed in 1932 when N. E. Brown transferred *M. pottsii* and '*M. crocosmiaeflora*' to *Crocosmia*. The original hybrid plant is still widely cultivated though it no longer compares favorably with newer hybrids with larger, longer-lasting flowers. Nevertheless, it remains one of the most reliable, free-flowering and persistent of the *Crocosmia* hybrids.

Crocosmia pottsii was collected repeatedly after it was formally named (as a *Montbretia*) in 1877. Collections across Natal, now officially the province of KwaZulu-Natal, at middle elevations from Kokstad in the south to Eshowe in the north, were made soon after that date. It is now known to be a common plant in Natal in streamside habitats at elevations up to 2000 feet (600 m).

Subsequent new discoveries in *Crocosmia* came rapidly after the 1890s. The important South African botanical figure Harry Bolus (1834–1911) discovered a new *Crocosmia* in 1896 in what was then called Tembuland. This territory corresponds to the interior of what is loosely called the Transkei, that is, the eastern half of Eastern Cape province, lying east of the Great Kei River. A second collection made at the same place, along the pass called Satan's (or Satana's) Neck in the mountains north of Engcobo, by the botanical artist Marianne H. Mason and her brother Edward Mason, principal of Saint Bede's College at Umtata, formed the basis for *Tritonia masoniorum* (at first spelled '*masonorum*').

Tritonia masoniorum was described in 1926 by the energetic Cape Town botanist Harriet Margaret Louisa Bolus (1877–1970), who also described another new species, *T. mathewsiana*, in 1921; both are now assigned to *Crocosmia. Crocosmia mathewsiana* was first collected in 1916 and grown to flowering at Kirstenbosch. A relatively rare endemic of the eastern escarpment of what was then the Transvaal, now Mpumalanga province, *C. mathewsiana* is confined to quartzite rocks and their immediate surroundings in the Graskop and Blyde River Canyon portion of the lower escarpment of the northern Drakensberg.

Crocosmia ×*crocosmoides*, the result of a cross between *C. aurea* and *C. paniculata*, first appeared in the horticultural literature in the van Tubergen catalogue of 1895 as *Antholyza crocosmoides*. Then a valid botanical description appeared in *Garden and Forest* (Gerard 1897). Quite independently in 1932, N. E. Brown named the same plant '*C. latifolia*' as if it were a wild plant. Evidently, *C.* ×*crocosmoides*, under the name *A. crocosmoides*, was sent to the Royal Botanic Gardens, Kew, by the van Tubergen nursery of Haarlem in 1904. Three herbarium specimens were supplied by the nursery's proprietor, Cornelius Gerrit van Tubergen. The first two (denoted as the type specimens) were in bud; a third, supplied some weeks later, had almost finished flowering. The specimens preserved in the Kew herbarium represent the type material of both *C.* ×*crocosmoides* and *C.* ×*latifolia* and appear to be the only physical record of the latter. Nothing resembling the plant has ever been found in the wild. Miriam de

Vos first recognized it (as *C. latifolia*) as a good species in 1984 but in 1999 regarded the plant as of hybrid origin. We concur and agree that it has no place in an account of the wild species of the genus; we refer to it throughout as *C. ×crocosmoides*.

The single species of *Crocosmia* that occurs outside the African continent is the Madagascan *C. ambongensis*, a plant of the western part of the island. Evidence suggests that it is a rare endemic of eroded limestone hills. It has been collected only twice, first in 1903 by the French explorer of the flora of Madagascar, H. Perrier de La Bâthie. It was described in 1939 when it was assigned to *Geissorhiza*, a genus otherwise restricted to the Cape region of South Africa. The plant lacks the characteristic woody corm tunics of *Geissorhiza* and its features seem most consistent with *Crocosmia*, to which it was transferred in 1990 by Goldblatt and Manning. The species rests there somewhat uncomfortably for it is a small plant with yellow flowers not particularly closely resembling the continental African species of the genus. The collection lacks mature capsules and seeds, important for distinguishing *Crocosmia* from the related *Tritonia*. Its position in *Crocosmia* depends largely on leaf anatomical evidence. Additional material is needed for further study, but for the present this unusual Madagascan plant remains in *Crocosmia*.

The last wild species of *Crocosmia* to be discovered is *C. pearsei*, brought to botanists' attention by the KwaZulu-Natal teacher and conservationist Reginald Oliver Pearse in his 1978 book, *Mountain Splendour*. This plant, until then confused with the related *C. paniculata*, is restricted to higher elevations of the Drakensberg, mostly above 7000 feet (2000 m), in Lesotho and adjacent parts of the South African provinces of KwaZulu-Natal and Free State. Its habitat—steep rocky slopes and cliffs—makes it particularly difficult to collect, but it is nevertheless surprising that it escaped notice until comparatively recently. Other equally rare and hard-to-collect Drakensberg species have been known for a century. The fabled suicide gladiolus, *Gladiolus flanaganii*, which grows on basalt cliffs near the elevation of and not far from populations of *C. pearsei*, was first collected in 1894. *Crocosmia pearsei* was formally described by the South African botanist Amelia A. Obermeyer in 1981.

Plant Form and Structure

Corm

Chasmanthe and *Crocosmia* have a corm, that is, an underground bulb-like storage organ derived from stem tissue. Like all members of subfamily Crocoideae of the family Iridaceae, the corm consists of a few internodes and produces roots from the lower portion of the organ. The corms have firm, dry, papery coverings called tunics that are derived from the bases of the lowermost leaves. At the end of each growing season, a new corm is produced at the base of the flowering stem, thus immediately on top of the current corm. The new corm has a terminal bud that will produce a new plant at the beginning of the next growing season. Lateral buds at the nodes of the corm may also produce new plants. Corms often also produce runner-like, slender rhizomes called stolons that grow some distance from the parent plant and then develop a small terminal plantlet with its own corm. This feature is particularly conspicuous in *Crocosmia aurea* (Plate 5) so that, in a few seasons, one plant can produce a large colony of apparently separate plants—all clones of one parent corm.

Corms of *Crocosmia* are persistent, so those of past seasons can often be seen attached in a chain to the base of the current corm. In *C. mathewsiana* (Plate 8) and *C. pearsei* (Plate 10) there are always numerous old corms of various sizes around the base of the plants, a distinctive feature that suggests a close relationship between the two despite their rather different flowers.

In contrast to those of *Crocosmia*, the corms of *Chasmanthe* are not persistent, and like most corms are almost completely resorbed annually by the plant during the course of the growing season, leaving at most a dry, leathery, flattened disk.

Stem and Leaves

Stems of *Chasmanthe* and *Crocosmia* are aerial and extend well above the ground. Round in cross section, they bear progressively smaller leaves higher up the stem. Stems may be simple, and they are consistently so in *Chasmanthe aethiopica* and usually so in *Crocosmia masoniorum*, or are normally branched. Branching is particularly pronounced in *Crocosmia fucata*, *C. mathewsiana* and *C. paniculata*, in which the branching pattern is distinctive. The branches diverge at a wide angle from the main stem, and the lowermost branches may branch again. The branches of these three species are crowded toward the upper third of the stem and form an almost compound inflorescence, sometimes called a pseudopanicle, that bears hundreds of flowers.

Leaves of *Chasmanthe* and *Crocosmia* conform in general to the basic morphology in the entire family Iridaceae. Thus the sword-like leaves have a sheathing base enveloping the lower part of the stem, and a unifacial blade oriented edgewise to the stem. The leaves overlap one another at their bases and form a two-ranked fan just as in, for example, *Iris* and *Gladiolus*. In subfamily Crocoideae, all members of which have a corm, the first leaves produced at the beginning of the growing season lack blades and serve as a protective sheath for the emerging leaf blades. These bladeless leaves are called cataphylls, and their bases become the tunics of the new corm to be developed as the season progresses.

Foliage leaves of *Chasmanthe* are distinctive and have a moderately firm texture, a pronounced central vein, and numerous tiny secondary veins set close together. The leaves are more or less sword-shaped and together form a loose fan. Basal leaves reach to about the base of the spike, or in *C. aethiopica* may slightly exceed the spike.

Right Leaf anatomy of *Crocosmia*. **A** Section through the leaf of *C. aurea*, showing the thick pseudomidrib and equally developed pairs of vascular bundles, ×40. **B** Detail of leaf margin of *C. aurea*, showing thickened, columnar epidermal cells, ×400. **C** Section through the leaf of *C. paniculata*, showing the pleated shape and unequally developed pairs of vascular bundles, ×35. **D** Section through the leaf of *C. masoniorum*, ×35. **E** Detail of leaf margin of *C. masoniorum*, showing thickened, columnar epidermal cells, ×350. Solid shading indicates sclerenchyma and xylem; hatching, phloem. Drawing by John Manning.

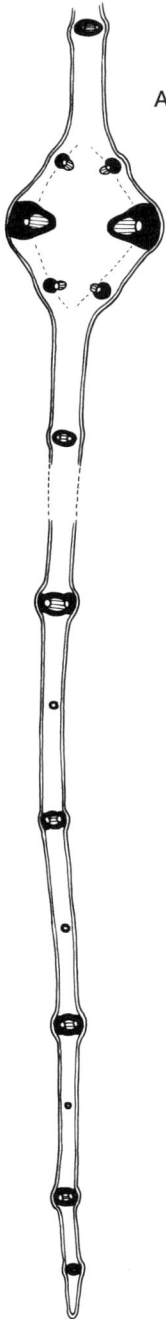
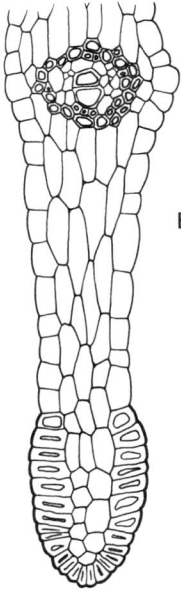
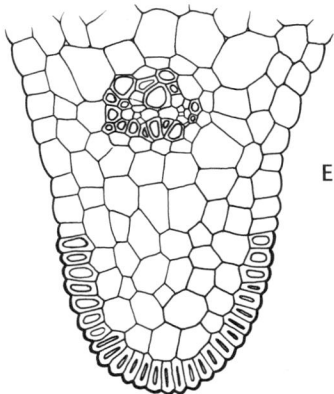
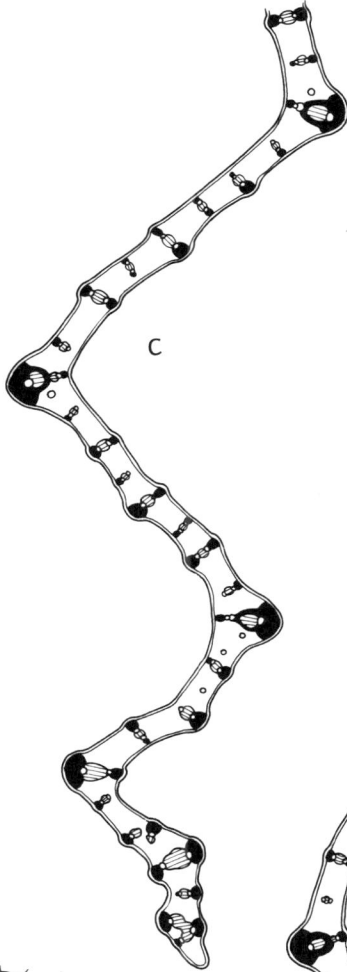

A

B

C

D

E

In *Crocosmia*, leaf morphology is more variable. Four species (*C. ambongensis*, *C. aurea*, *C. fucata* and *C. pottsii*) have sword-shaped, plane leaves with a pronounced central vein. The other veins are inconspicuous. The leaves of *C.* ×*crocosmiiflora* (*C. aurea* × *C. pottsii*) have this morphology. The four other species (*C. masoniorum*, *C. mathewsiana*, *C. paniculata* and *C. pearsei*) have broader, lanceolate leaves with pleated blades, and multiple primary veins of more or less equal size.

Anatomically, leaves of *Chasmanthe* and *Crocosmia*, including those species of the latter that have pleated blades, have two important features (de Vos 1982b, Rudall and Goldblatt 1991, Rudall 1995). The leaf margins have columnar epidermal cells with the walls at 90° to the surface thickened, and the margins also a lack marginal vein and the strand of sclerenchyma (cells with heavily thickened walls) beneath the epidermis that are found in many Iridaceae. These two anatomical features are specializations that occur in several genera of subfamily Crocoideae, including *Tritonia* as well as *Devia*, *Duthieastrum*, *Freesia* and *Sparaxis*. The central leaf veins of *Chasmanthe* and *Crocosmia* species with plane leaves always consist of multiple vascular strands. In related genera, the central vein consists of just a pair of opposed vascular strands.

Table 1. Orientation of styloid crystals in leaves of *Chasmanthe*, *Crocosmia* and related genera*

Species	Crystal orientation
Chasmanthe aethiopica	random
Crocosmia ambongensis	random
Crocosmia aurea	random
Sparaxis bulbifera	parallel
Tritonia crocata	parallel
Tritonia pallida	parallel
Tritonia securigera	parallel

* Data are from Goldblatt and Manning (1990a) with additions not before published. Orientation is not known in *Chasmanthe bicolor*, *C. floribunda*, *Crocosmia fucata*, *C. masoniorum*, *C. masoniorum*, *C. mathewsiana*, *C. paniculata*, *C. pearsei* and *C. pottsii*.

De Vos (1982b, 1984), who made a detailed anatomical examination of *Chasmanthe, Crocosmia* and *Tritonia,* also found that *Crocosmia* species may be distinguished by the way the secondary veins originate. In *Tritonia* they are present in the sheath and run directly into the blade whereas in *Crocosmia* they branch off the primary central vein above the sheath, thus within the blade itself, a feature confirmed for the Madagascan *C. ambongensis* (Goldblatt and Manning 1990a). Another peculiar feature of *Crocosmia* leaves is the random orientation of the calcium oxalate crystals in the blades; the crystals lie parallel to the veins that traverse the long axis of the leaf in *Tritonia.* Examination of additional species (Table 1), including *Chasmanthe,* confirms the distinction between *Crocosmia* and *Tritonia,* and we now know that *Chasmanthe* shares the random crystal orientation of *Crocosmia.* The significance of this character is uncertain.

Inflorescence

Like nearly all genera of Iridaceae subfamily Crocoideae, the inflorescence of *Chasmanthe* and *Crocosmia* is a spike, that is, the flowers are sessile on an elongate axis. Flowering stems bearing the spikes may be simple or branched in various ways, sometimes so densely that the inflorescence appears to be a compound structure, sometimes called a pseudopanicle (from the botanical term panicle, a multibranched inflorescence, and the Latin *pseudo-,* false). The number and arrangement of the flowers on a spike is important, and main spikes always have more flowers than the lateral ones (or branches). As many as 30–40 flowers per spike are characteristic of *Chasmanthe floribunda,* and well-grown flowering stems of *Crocosmia paniculata* may bear 150–200 flowers, with as many as 25 on each branch.

Spikes are erect and have flowers in two opposed ranks in *Chasmanthe bicolor* and *C. floribunda,* but in *C. aethiopica* the spike is inclined toward the horizontal and the flowers are borne on the upper surface in a single crowded row. In *Crocosmia* the individual spikes are inclined, to more or less horizontal, but the flowers may be borne in two opposed rows, facing away from one another (*C. fucata, C. paniculata* and *C. pearsei*), or in more or less a single row (*C. masoniorum*). They are also carried in a single row in *C. ambongensis*, which has only two or three flowers, facing more or less the same direction.

Floral bracts, an important generic feature in subfamily Crocoideae, are set at the base of the sessile ovary. Both the outer (abaxial) bract and the inner (adaxial) one are relatively short and about equal in length. The single-veined outer bract is, however, somewhat broader and obtuse, or subacute or obscurely three-forked, whereas the inner bract is always two-veined and forked apically for a short distance. In *Crocosmia* the bracts are more or less leathery and at first green or flushed with purple, but they become dry, starting at the tips, with age and are usually partly dry when the flowers open. The bracts of *Chasmanthe* have a softer texture and may remain green after the flowers fade.

Flowers

Flowers of *Chasmanthe* and *Crocosmia* show similarities that understandably have led to the belief that the genera are closely related. It has even been suggested that they may be better treated as one genus. The shared floral similarities, including the elongate perianth tube, narrow below and wide and cylindric above, the bright orange to scarlet pigmentation and the well-exserted stamens and style, are now understood to be adaptations to pollination by sunbirds, however. This suite of characteristics do not, in themselves, indicate close relationship. Similar flowers occur in other genera of the *Iris* family, including *Gladiolus*, *Tritoniopsis* and *Watsonia*. In none of these is this the predominant or sole flower type, as it is in *Chasmanthe*. Vegetative structure and the nature of the fruit and seeds, and chromosome morphology in the African Iridaceae, are equally or more important in assessing relationship. Thus the genus *Antholyza*, established by Linnaeus in the 18th century for members of the *Iris* family with red or orange flowers with an elongate peri-

Right Flowers of *Chasmanthe* and *Crocosmia*. *Crocosmia mathewsiana*, **A** front view, **B** side view, **C** half-flower. *Chasmanthe aethiopica*, **D** front view, **E** side view, **F** half-flower. *Crocosmia fucata*, **G** front view, **H** side view, **I** half-flower. *Crocosmia aurea*, **J** front view, **K** half-flower, **L** outer (left) and inner (right) bracts, **M** longitudinal (left) and transverse (right) sections through ovary, **N** front (left) and side (right) views of anther, **O** style branches, **P** capsules, **Q** seed. All drawings ×0.8 except **M**, **N**, **O** and **Q**, which are variously enlarged. Drawing by John Manning.

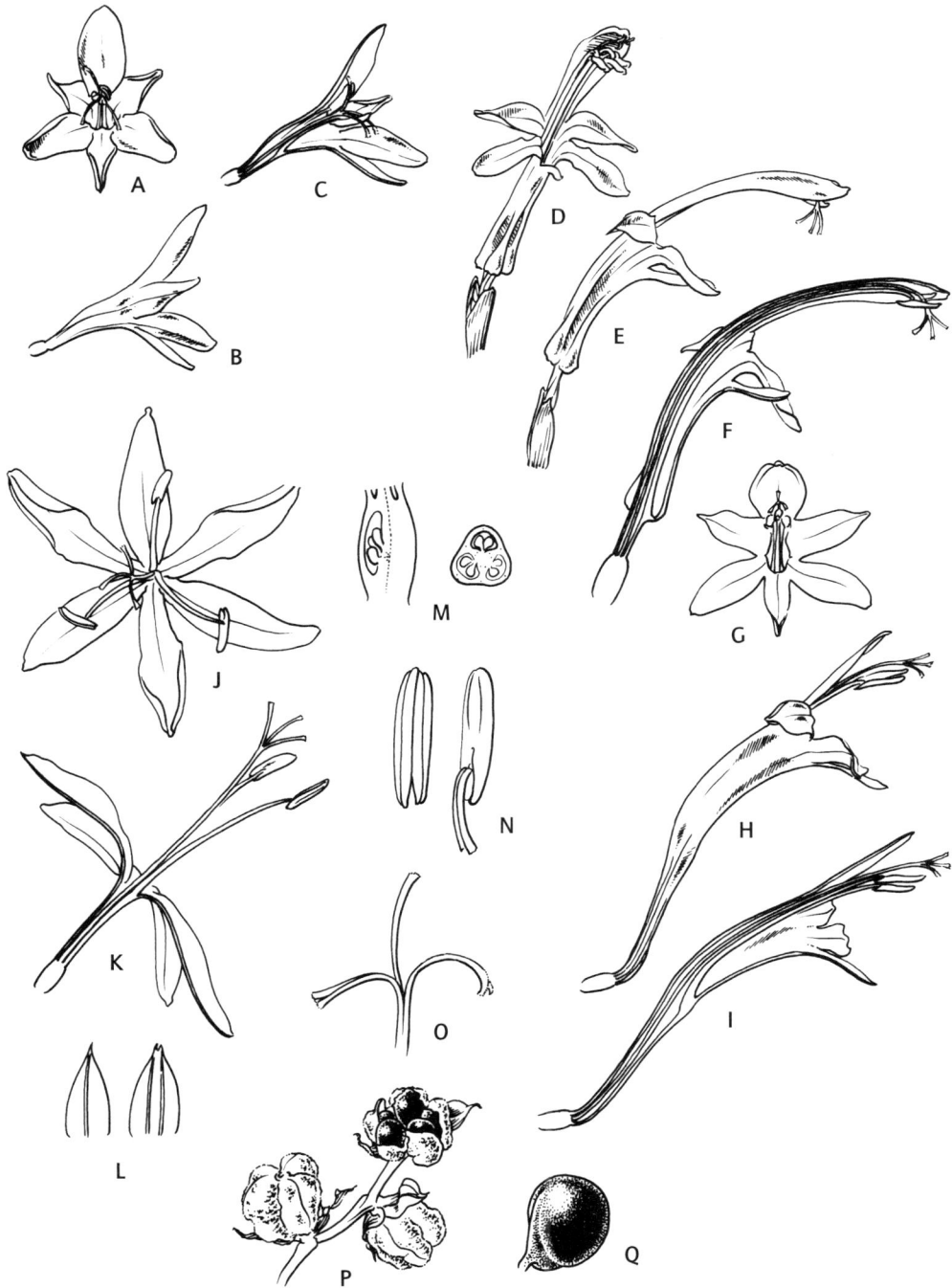

A B C D E F G H I J K L M N O P Q

anth tube, is now understood to be a heterogeneous assemblage of species that included species of *Babiana, Gladiolus, Tritoniopsis* and *Watsonia* as well as *Chasmanthe* and *Crocosmia*. Accordingly, our discussion of floral morphology of both genera does not assume relationship, merely convenience. In fact, while all *Chasmanthe* species have orange-scarlet flowers with elongate perianth tubes, some *Crocosmia* species have rather different flowers. The type of the genus, *C. aurea,* has nodding flowers with a more or less uniformly narrow tube, and it is the only species in the two genera that has radially symmetric flowers with tepals about the same size. Two other species of *Crocosmia, C. mathewsiana* and *C. pottsii,* have short-tubed flowers with short, arching stamens included in the perianth. These floral differences actually indicate adaptations to pollinators other than sunbirds. In species with elongate tubes, the dorsal tepal is significantly larger, and in *Chasmanthe* the dorsal tepal is narrow and spathulate while in *Crocosmia* the dorsal tepal is more or less triangular-deltoid.

Important floral features of *Chasmanthe* are the abrupt widening of the perianth tube at the top of the narrow lower part of the tube, not found in *Crocosmia*. The difference in the way the floral tube is constructed suggests that adaptation for sunbird pollination may in fact be the result of convergence. That is, the similarity of form evolved independently from different ancestors. The stamens of *Chasmanthe* and *Crocosmia* also differ in significant ways. In *Chasmanthe* the filaments are slightly unequal in length; the lower abaxial filament, which arches upward to lie between the two adaxial filaments, is slightly longer. As a result, the anthers, which are equal in length, reach different levels, the median one protruding beyond the tips of the other two. In *Crocosmia* and other bird-pollinated Iridaceae, the stamens have the same orientation but are equal in length so that the anther tips reach the same point. The anthers are versatile, that is, able to swing back and forth on the top of the filament, to which they are attached just below the middle. They are also distinctive in having the lower halves slightly diverging, for about half their length in *Chasmanthe* but for only about one-third of their length in *Crocosmia.*

Pollen grains of *Chasmanthe* and *Crocosmia* are to all appearances identical, and exactly like those of *Tritonia* and most other genera of subfamily Crocoideae (Goldblatt et al. 1991). The grains have a thin exine, the outer layer of which has minute perforations, and the surface bears scattered short excrescences. Technically, this exine is called perforate-scabrate. The pollen grain has a broad aperture (or sulcus), an area of the grain lacking an exine layer, running the length of the long axis, through which the pollen tube emerges. The smooth aperture membrane has a pair of distinctive, narrow, longitudinal exine bands running almost the length of the grain. These bands, unique in the Iridaceae to the majority of genera of Crocoideae, constitute what is usually called an operculum.

De Vos (1984) also pointed out that the style branches of *Crocosmia* are short, and slightly expanded and minutely forked at the apex. Style branches of *Chasmanthe* are similar but not notched apically. Curiously, in both genera the style branches are stigmatic only at the tips. In contrast, the style branches of the related genus *Tritonia* are longer, uniformly narrow and stigmatic along part of their length.

Fruit and Seeds

The fruit of both *Chasmanthe* and *Crocosmia* is a capsule, that is, a dry fruit with firm to bony, dry walls that split apart in the middle of the locules or seed chambers when ripe to expose multiple seeds in three separate locules (illustrated with the drawings of the flowers), as in all members of the Iridaceae. Poor fertilization sometimes results in one or even two of the locules being empty of seeds and then smaller than those that bear seeds. Capsule structure in the two genera is unremarkable but nevertheless provides some useful characters for understanding species and their relationships. Capsules are more or less ovoid and three-lobed in *Chasmanthe* and show the outline of the seeds. In *C. floribunda* the capsules taper at the tip to a distinctive, nipple-like apex. *Crocosmia* capsules are similar but wider than high, except in *C. pottsii*, which has more or less globose capsules, the walls uneven and showing the outlines of the seeds. The capsules of some *Crocosmia* species, notably *C. aurea, C. fucata* and *C. pearsei,* have a slightly rough, wrinkled and papillate surface that recalls the capsule surface of *Freesia*.

Seeds show more important differences between the genera. As in all seed plants, the seeds of *Chasmanthe* and *Crocosmia* are ripened, fertilized ovules. In most flowering plants each ovule is enclosed by two distinct layers or coats, termed integuments, which develop into the seed coat. In Iridaceae the inner integument is usually crushed during ripening of the seed, and it is only the outer integument that contributes significantly to the mature seed coat.

Chasmanthe species have globose, bright orange, unusually large seeds, as much as ¼ inch (7 mm) in diameter in *C. aethiopica*. In this species the seed coat in fresh seeds is fleshy and sweet to the taste. *Chasmanthe bicolor* and *C. floribunda* also have orange seeds but with a hard, smooth coat. The orange pigmentation is contained in the large, thick-walled cells of the epidermis according to de Vos (1985). The bright color is probably an adaptation to bird dispersal. The seeds of these two species are deceptive, for collectively in the open capsule they resemble a small fleshy berry. Fleshy fruits are, of course, attractive to fruit-eating birds, but except in *Chasmanthe aethiopica* the seeds offer no food reward. Only the fleshy seed coat of *C. aethiopica* provides a food reward to fruit-eating birds, which in turn provide a service to the plant: dispersal of the seeds. Thin sections of the developing seed coat of *C. aethiopica* show that there are seven layers of large, water-filled parenchyma cells beneath the epidermis, compared with the usual three of four layers of smaller cells in other genera of the family. Offering no edible reward, the seeds of *Chasmanthe bicolor* and *C. floribunda* appear to function like the hard, bright orange seeds of the legume genus *Erythrina* (lucky bean tree), attracting birds to eat them but providing no nutrition and passing through the bird's digestive tract undamaged, to be excreted some distance from the parent plant.

The seeds of the species of *Crocosmia* are globose or angular, dry, and brownish, reddish or black when mature. They have a rough surface with the cell outlines clearly visible, and they are lightly wrinkled. In *C. aurea* the seed coat is smooth and slightly moist when released, but later dries and becomes wrinkled. The black pigmentation of these seeds is developed in the cells of the epidermis, where it can be seen as numerous small

Right Seed development in *Crocosmia aurea*. During ripening of the seed, the outer integument (*oi*, termed the testa at maturity) develops into the seed coat while the inner integument (*ii*, termed the tegmen at maturity) is crushed. **A** Longitudinal section through ovule, ×85. **B** Transverse section through ovule, ×85. **C** Detail of integuments, ×350. **D** Longitudinal section through seed, ×20. **E** Detail of seed coat, ×85. **F** Transverse section though vascular region of seed (cellular detail of the vascular bundle omitted), ×85. The vascular bundle is indicated in **A**, **B** and **D** by the broken border. Drawing by John Manning.

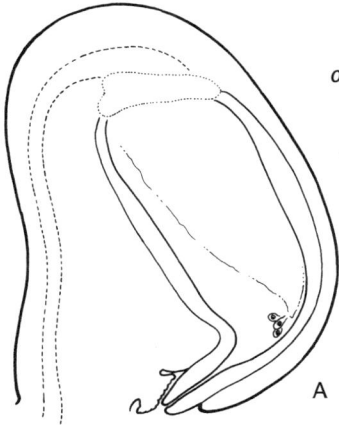

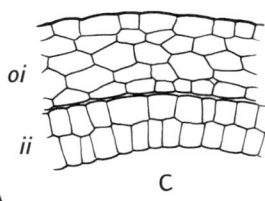

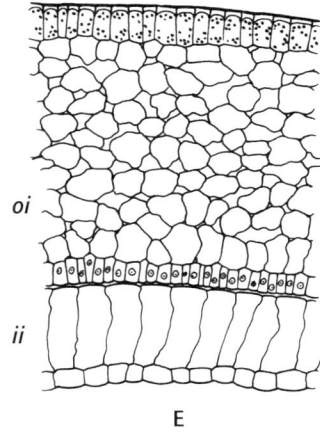

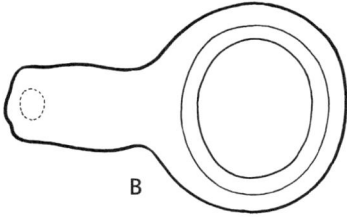

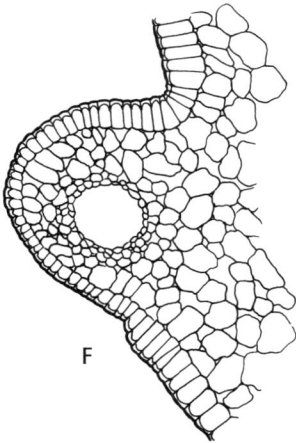

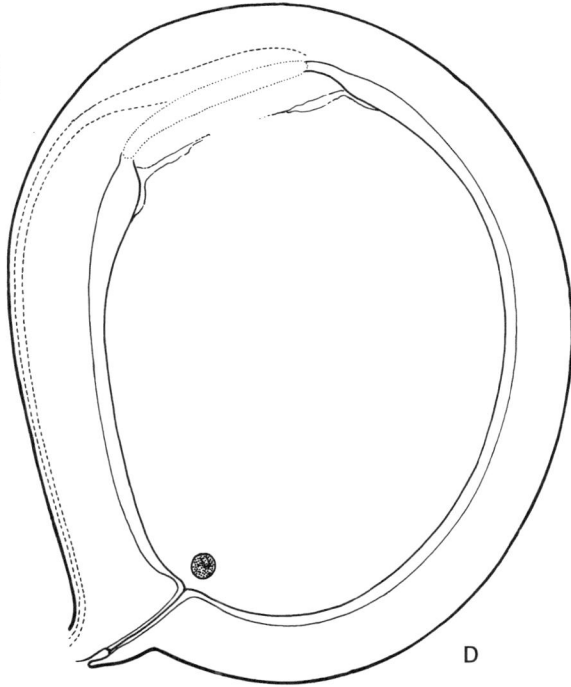

A

oi

ii

C

oi

ii

E

B

F

D

dark granules under the microscope. Thin sections of the developing seed coat show about eight layers of large, water-filled parenchyma cells beneath the epidermis, each containing numerous starch grains. As in *Chasmanthe aethiopica,* this thick, spongy layer contrasts with the usual three or four layers of smaller cells typically found in the seed coats of other species. The extra layers are formed by cell division from the three cell layers present in the ovule. De Vos (1984) described the seeds of *Crocosmia aurea* as having a layer of large, thin-walled, water-retaining cells under the epidermis. As the seeds dry, this layer breaks down and the epidermis is easily rubbed off, exposing a dull-colored, wrinkled inner layer. The inside of the capsules of *C. aurea* is usually flushed reddish or orange, making a bright background for the nearly black seeds. We assume again, as in *Chasmanthe bicolor* and *C. floribunda,* that the fruit resembles a fleshy berry, designed to attract fruit-eating birds to feed on them while actually offering little or possibly even no reward. The seeds of other *Crocosmia* species are dark reddish brown except in *C. pottsii,* and the capsules walls are straw-colored. There is no evidence that they function in the same way as those of *C. aurea,* to attract birds to disperse the seeds. The seeds of *C. pottsii* are unusual in the genus in their light brown color, smaller size and spongy seed coat. Unlike seeds of other *Crocosmia* species, they are evidently adapted for dispersal by water, appropriate for a species that grows along stream banks. The distinctive seed features that are developed in *C. aurea* appear to be a specialization within the genus rather than a generic character of *Crocosmia.*

Chasmanthe and Crocosmia seeds have an interesting feature discovered during an examination of seed development in the Iridaceae (Goldblatt and Manning 1995). In *Chasmanthe* the epidermis overlying the vascular trace that supplies the ovule with nutrients does not develop the thickened walls that are formed in the surrounding epidermal cells. As a result, the vascular trace becomes excluded from the seed rather than embedded in the seed coat, as is normal, and comes to lie outside the seed body as a minute thread attached to the chalazal end of the seed. It is easily torn free, and although in older seeds the vascular trace may be lost, the seed still shows the scar along the ridge leading from the base of the seed to the chalaza. The excluded vascular trace, considered to be a significant trait for understanding the relationships of genera of the subfamily Crocoideae, is shared with *Tritonia* and the related *Devia, Duthieastrum* and *Sparaxis* as well as the apparently more distantly related genera *Dierama* and *Ixia*. Thus only 8 of the 28 genera in Crocoideae have these specialized seeds.

Although the seeds of *Crocosmia* appear to share this characteristic, careful examination of the seed coat suggests that it is arrived at in a slightly different way. In *Crocosmia* the vascular trace lies in a small ridge along the outside of the seed body, and the epidermal cells that cover it, though significantly smaller than the surrounding cells, develop a thickened outer wall. The underlying tissue of the seed coat is spongy and easily torn, so the vascular trace is also easily stripped from the seed. The evident similarity in the excluded vascular trace in seeds of *Chasmanthe* and *Crocosmia* thus appears to have been arrived at in different ways, and the excluded traces may well have evolved independently in the two genera.

Seed number is significant in both *Chasmanthe* and *Crocosmia* and was one of the characters used by de Vos (1984) to separate the two genera from *Tritonia*. The somewhat woody capsules of *Crocosmia* have as many as three seeds per locule, occasionally four, according to de Vos, whereas the softer-walled capsules of *Tritonia* contain many round, hard-walled seeds per locule. De Vos's distinction fails when the capsules of *Crocosmia pottsii*, a species unknown to her, are taken into account. They contain about 24 small angular seeds with a slightly spongy coat, and the capsules themselves are more or less globose. Thus, as regards capsule shape and seed number, *C. pottsii* is more like *Tritonia*, but the seed surface of *C. pottsii* reveals the cell outlines and has a matte texture, which is different from that of *Tritonia*, which has seeds with a hard shiny coat.

Chromosomes

Chromosome number is an important taxonomic character at the level of genus in *Crocosmia* and *Chasmanthe*. The latter has a basic chromosome number $x = 10$, and the two species counted (Table 2) are diploid, with a somatic chromosome number $2n = 20$ (Goldblatt 1971). Other numbers published in the literature may be miscounts or based on misidentifications. Chromosomes of *Chasmanthe* show no significant features except for their small size and relatively uniform shape.

All *Crocosmia* species counted have the basic chromosome number $x =$

11, and all species known (Table 2) are diploid, with a somatic chromosome number $2n = 22$ (Goldblatt 1971, de Vos 1984). Like *Chasmanthe*, the chromosomes of *Crocosmia* are small and more or less uniform in size.

The difference in chromosome number between *Chasmanthe* and *Crocosmia* provides some useful evidence in support of maintaining them as separate genera. The larger genus *Tritonia* (with about 28 species), in the past believed to be closely allied to *Chasmanthe* and *Crocosmia*, has a basic chromosome number $x = 11$. This suggests a closer relationship to *Crocosmia* and casts some doubt on the presumed close relationship between *Chasmanthe* and *Tritonia*.

Table 2. Chromosome numbers of *Chasmanthe* and *Crocosmia**

Species	Diploid chromosome number $2n$
Chasmanthe aethiopica	20
Chasmanthe floribunda	20
Crocosmia aurea	22
Crocosmia ×*crocosmiiflora*	22
Crocosmia fucata	22
Crocosmia masoniorum	22
Crocosmia paniculata	22
Crocosmia pottsii	22

* Data are from Goldblatt (1971) and de Vos (1984, 1985). Counts are not available for *Chasmanthe bicolor, Crocosmia ambongensis, C. mathewsiana* and *C. pearsei;* a report by Goldblatt (1971) for *C. mathewsiana* was shown by de Vos (1984) to be based on misidentified plants that are correctly *C. paniculata*.

Ecology

Pollination

The pollination of *Chasmanthe* and *Crocosmia* has not been much studied, and most of the observations here are original. The flowers of all three species of *Chasmanthe* have the characteristics of other African Iridaceae, and of plants in general, that are pollinated by sunbirds, species of *Nectarinia* (Goldblatt et al. 1999). They are orange to red, sometimes with green or yellow markings, and have an elongate tube, relatively wide and more or less tubular in the upper half and slender in the basal half. These flowers produce comparatively large quantities of nectar from septal nectaries, that is, nectar-secreting glands in the radial walls (septa) of the ovary. Nectar is secreted via small pores in the top of the ovary directly into the perianth tube, where is accumulates when the flowers are open. Nectar volumes of as much as 30 μl (microliters) have been recorded in *Chasmanthe aethiopica*. In general, nectars that flowers offer to birds have a relatively low nectar concentration, though there are some important exceptions, even within the family Iridaceae. *Chasmanthe* species, however, do conform to this trend (Table 3). Nectar concentrations of 10% sugar have been reported for *C. bicolor*, and 15.7–17% sugar for *C. aethiopica* and *C. floribunda*, levels that are consistent with bird pollination. Nectar sugar chemistry also reflects the bird pollination in these species, for the nectar of *Chasmanthe* flowers has high proportions of the so-called pentose sugars (glucose and fructose) and low levels of the hexose sugar (sucrose). Species of subfamily Crocoideae that are pollinated by bees, moths and long-proboscid flies invariably have nectars with higher sucrose concentrations: so-called hexose-rich or sucrose-dominant nectars in the terminology of Baker and Baker (1983).

Comparable nectar characteristics have been obtained for two bird-pollinated *Crocosmia* species for which we have nectar data, *C. fucata* and *C. paniculata*. Flowers of *C. fucata* produce as much as 11.5 μl of nectar with an average concentration of 18.0% sugar, figures closely matched by those of *C. paniculata,* which has a nectar concentration of 17.4%. Analysis of *C. fucata* nectar also shows unusually high levels of glucose and fructose (pentose sugars) and a sucrose : pentose ratio of 0.53, which is regarded as pentose rich (Baker and Baker 1983), a pattern consistent with bird pollination.

Table 3. Nectar characteristics of flowers of *Chasmanthe* and *Crocosmia* *

Species	Nectar volume μl (*n*)	Nectar concentration % (s.d.)	Range of sugar %			Ratio sucrose : (fructose + glucose) (*n*)
			Fructose	Glucose	Sucrose	
Chasmanthe aethiopica	24–30 (7)	15.7 (1.61)	43–45	40–42	14–15	0.17 (3)
Chasmanthe bicolor	11–12 (2)	10.0 (n.a.)	45	52	3	0.03 (1)
Chasmanthe floribunda						
population 1†	6.4–7.2 (3)	15.7 (0.28)	45–46	46–53	2–9	0.06 (3)
population 2	17–19 (2)	17.3 (n.a.)	46	48	6	0.07 (1)
Crocosmia aurea						
population 1	2.8–3.4 (3)	24.3 (0.58)	n.a.	n.a.	n.a.	n.a.
population 2	2.3–3.7 (5)	17.7 (1.04)	n.a.	n.a.	n.a.	n.a.
population 3	1.9–2.8 (5)	21.8 (0.06)	n.a.	n.a.	n.a.	n.a.
Crocosmia fucata	9.8–11.5 (3)	18.0 (2.78)	20–34	28–40	26–52	0.53 (3)
Crocosmia paniculata	11.5–15.8 (7)	17.4 (1.02)	n.a.	n.a.	n.a.	n.a.
Crocosmia pottsii	<1.0 (3)	23.2 (2.25)	n.a.	n.a.	n.a.	n.a.

* Note the large differences in volume between flowers adapted for pollination by sunbirds (*Chasmanthe* spp., *Crocosmia fucata* and *C. paniculata*) and those pollinated by bees or butterflies (*Crocosmia aurea* and *C. pottsii*). Data for *Chasmanthe* and *Crocosmia fucata* are from Goldblatt et al. (1999). *n,* Number of samples; μl, microliters; n.a., not available; s.d., standard deviation.

† The low score for nectar volume in population 1 of *Chasmanthe floribunda* indicates that sunbirds had already taken substantial amounts of nectar from these flowers.

The only other species of *Crocosmia* for which we have nectar data are *C. aurea* and *C. pottsii*. The nodding flowers or *C. aurea*, evidently adapted for pollination by large *Papilio* butterflies, produce as much as 3.7 µl of nectar with average sugar concentrations for different population samples of 17.7–24.3%. The flowers of *C. pottsii*, which appear to be adapted for pollination by bees, have an average sugar concentration of 23.2%. No data are available for nectar sugar chemistry. The relatively low sugar concentration in *C. aurea* nectar is typical of butterfly-pollinated flowers.

Chasmanthe species have long been considered to be pollinated by sunbirds, and the first such published observations date from the late 19th century when the pioneer pollination biologist George Scott Elliot (1890) recorded visits by the lesser double-collared sunbird, *Nectarinia chalybea*, to *C. aethiopica*. The same species of sunbird was observed visiting *C. aethiopica* by Vogel (1954) and Goldblatt et al. (1999), the latter also reporting visits by the malachite sunbird, *N. famosa*, to a second population. Visits by the lesser double-collared sunbird to *C. floribunda* have also been recorded.

Pollination observations for a few *Crocosmia* species by sunbirds are available (Goldblatt et al. 1999), and not surprisingly these few species have flowers very much like those of *Chasmanthe* species. In the Drakensberg, the malachite sunbird has been noted visiting *Crocosmia pearsei*, and the double-collared sunbird, *Nectarinia afra*, has been observed visiting *C. paniculata* near Graskop in Mpumalanga province. The Namaqualand *C. fucata* is also pollinated by malachite sunbirds.

The only known visitor to the narrow-tubed flowers of *Crocosmia aurea* is the large swallowtail butterfly, *Papilio nireus*. Examination of *C. aurea* flowers shows that only insects with long, slender mouthparts can reach the nectar because the bases of the three filaments partially block the mouth of the tube, leaving three tiny channels between the filament bases that permit entry to the tube, which may be as long as $1^{1}/_{16}$ inches (27 mm). A butterfly hangs upside down on the nodding flowers and inserts its slender proboscis into the tube while grasping the tepals with its legs. We suspect that flowers of *C. masoniorum* are likewise pollinated by large butterflies, for like *C. aurea* it has a relatively short, narrow perianth tube, spreading tepals and well-exserted stamens, thus broadly resembling those of *C. aurea* except that they are held erect instead of nodding.

Lastly, the flowers of *Crocosmia mathewsiana* are adapted for pollination by anthophorine bees, which are large, fast-flying and have very long tongues. Near Graskop we noted a species of *Amegilla* (family Apidae) visiting the plant, the dorsal part of the thoraxes of the insects sprinkled with the pale creamy pollen of *C. mathewsiana* flowers. We assume the Madagascan *C. ambongensis* as well as *C. pottsii,* which has small, short-tubed flowers, are also pollinated by large nectar-drinking bees. Thus *Crocosmia,* unlike *Chasmanthe,* displays adaptations for pollination by a variety of animals, including birds, large bees and butterflies, but all the species are specialists, having pollinators of only one class, and sometimes only one species.

The adaptations for bird pollination in *Crocosmia* appear to function as well outside their natural range as in their native habitat. The *Crocosmia* cultivar 'Lucifer' (Plate 13), now widely cultivated in northwestern North America, has flowers typical of bird-pollinated species though it is actually a cross between bird-pollinated *C. paniculata* and the interspecific hybrid *C. ×crocosmiiflora,* one of the parents of which has flowers adapted for bee pollination and the other for butterfly pollination. Nevertheless, the floral characteristics for bird pollination in 'Lucifer' seem dominant, and the flowers are visited by hummingbirds, the most important bird pollinators in North and South America. Hummingbirds do not occur in Africa, where avian pollinators are sunbirds (family Nectarinidae), which are unrelated to hummingbirds (family Trochilidae).

Plant Geography

Both *Chasmanthe* and *Crocosmia* are African genera. *Chasmanthe* has the narrower range, restricted to the southern and western coasts of South Africa. All three of its species favor a winter-rainfall regime, but *C. aethiopica,* which extends farthest to the east, reaches Kentani, a short distance east of East London, which lies in an area that receives ample rainfall in summer. *Chasmanthe bicolor* has the narrowest range, occurring locally in river valleys east of Cape Town at the foot of the Langeberg and in the area known as the Robertson Karoo. *Chasmanthe floribunda* has the widest range, with its northernmost station in coastal central Namaqualand near the hamlet of Kotzesrus. It extends south from there along the

coast to Hermanus, east of Cape Town, and locally inland on the Bokkeveld Plateau and in the adjacent mountains. All three species favor light
bush or margins of coastal forest, *C. aethiopica* typically growing under
shrubs and small trees. Following the rainfall pattern, *Chasmanthe* species
are winter-growing plants that begin their growth cycle with the first autumn rainfall and die back in late spring as the soil dries out in warmer
weather as summer approaches. *Chasmanthe aethiopica* flowers rapidly
after new growth begins, most often in late autumn (May or June). Last to
bloom is *C. floribunda,* in late winter and early spring (August and September).

In contrast, *Crocosmia* species occur widely across sub-Saharan Africa,
and one, *C. ambongensis,* occurs on the large Indian Ocean island of
Madagascar. Only one species, the localized *C. fucata,* occurs in the small
southern African winter-rainfall zone, where it grows along streams in
the mountainous Kamiesberg of Namaqualand, interior to the western
coast. The remaining species are summer-growing and dormant in the
dry season, which falls in the winter months. Of these, the most widespread species is *C. aurea,* a forest and forest-margin plant that extends
from Uganda, northern Congo and Central African Republic north of the
equator, across central Africa to the southeastern coast of South Africa.
Plants are most often seen in the wild in forest margins or clearings, but
they seem to grow equally well in the deep shade of evergreen forest.

Crocosmia paniculata has the next widest range. Centered in Mpumalanga province of South Africa, it extends south through Swaziland to the
foothills of the mountainous Drakensberg in KwaZulu-Natal, always
growing in wet grassland or seeps at the headwaters of streams. It is absent
from Limpopo province of South Africa but also occurs in eastern Zimbabwe, though we doubt it is native there. Closely related to *C. paniculata,*
C. pearsei is native to the high Drakensberg of Lesotho and adjacent parts
of KwaZulu-Natal and Free State provinces of South Africa, where it grows
on basalt cliffs and rock outcrops above 7000 feet (2000 m). Also favoring
rocky habitats and cliffs, *C. masoniorum* grows along a short stretch of the
southern Drakensberg in South Africa's Eastern Cape province.

Like *Crocosmia aurea, C. pottsii* favors light bush but only along steams,
its corms growing in permanently wet ground with the stems usually
arching gracefully over low banks above the water. Lastly, *C. mathewsiana,*

a very local species of central Mpumalanga, is restricted to the mist belt along the edge of the steep escarpment near the town of Graskop. It grows in elfin forest or in open ground, always in well-drained sandy soil derived from quartzitic bedrock.

Although *Crocosmia fucata* is winter-growing, it flowers in early summer and is dormant in late summer. The remaining species of the genus grow during the summer months, initiating growth in October or early November and flowering from midsummer onward. First to flower is *C. pearsei*, several weeks before *C. masoniorum* in late December and January, soon followed by *C. paniculata* and *C. pottsii*. *Crocosmia aurea* and *C. mathewsiana* rarely flower before late summer (mid-February) and last into mid-autumn (April). North of the equator, *C. aurea* flowers in June, July, August and well into September—depending on the climate or weather in a particular year.

Evolution and Classification

As mentioned at various points in the preceding pages, there is a prevailing belief that *Chasmanthe* and *Crocosmia* are closely allied, related in some uncertain way to the larger and to some extent, less-specialized genus *Tritonia*. Similarities in the flowers of *Chasmanthe* and some species of *Crocosmia* show how that belief developed. The predominant orange or red pigmentation of the flowers, specialized leaf anatomy (at least in species with plane leaf blades), unspecialized short floral bracts, undivided or apically notched or forked style branches and, in the case of *Crocosmia* and *Tritonia*, chromosome number $2n = 22$ constitute the evidence for relationship.

More generally, however, it seems beyond dispute that *Chasmanthe* and *Crocosmia* are part of a wider group of genera belonging to subfamily Crocoideae (Ixioideae, a better-known name for the subfamily, is a synonym). Members of the subfamily, which includes such well-known genera as *Gladiolus, Ixia, Sparaxis* and *Watsonia,* have an inflorescence that is unique in the Iridaceae, a spike in which the sessile flowers are subtended by a pair of opposed floral bracts, and the flowers consistently have a perianth tube, which is not always present in other subfamilies of the *Iris* family. Flowers of most Crocoideae are also zygomorphic (bilaterally symmetric) or secondarily radially symmetric. In addition, genera of Crocoideae have a specialized rootstock, a corm, which has internal vasculature and produces roots from the base, and pollen grains, unique in the Iridaceae, that have a perforate exine and an operculum consisting of two longitudinal bands across the aperture.

Within Crocoideae, *Chasmanthe* and *Crocosmia* have a striking seed specialization, an excluded vascular strand (Goldblatt and Manning

1995). This remarkable seed character is shared with just a handful of other genera, including *Devia* (1 sp.), *Dierama* (about 44 spp.), *Duthieastrum* (1 sp.), *Ixia* (about 50 spp.), *Sparaxis* (15 spp.) and *Tritonia* (28 spp.). It suggests that these genera constitute a clade within which *Dierama* and *Ixia,* which have conventional leaf anatomy (an unmodified marginal epidermis and a well-developed subepidermal marginal strand of sclerenchyma), seem isolated. They may be ancestral to the remaining genera, which lack a strand of sclerenchyma beneath the marginal epidermis and instead have the specialized columnar marginal epidermal cells with conspicuously thickened walls. This argument leaves us with slightly more candidates for the immediate allies of *Chasmanthe* and *Crocosmia* than just *Tritonia*. Nevertheless, *Tritonia* seemed the most likely ancestor, until DNA sequence studies were undertaken. Other genera with excluded seed vasculature, including *Sparaxis* (which today includes *Synnotia*), have specialized, dry, papery and crinkled floral bracts, and most of the species have flowers of different pigmentation. *Devia* and *Duthieastrum,* genera with only a single species each, are so specialized that any immediate relationship to *Chasmanthe* and *Crocosmia* seems not only unlikely but would tell us little about their wider relationships.

It is useful here to repeat the differences between *Chasmanthe* and *Crocosmia*. One important feature is the shape of the perianth tube. In all species of *Chasmanthe* the tube is elongate, with a narrow basal part abruptly expanded into a wider upper part at the point where the filaments are inserted, whereas in *Crocosmia* the tube always expands gradually, even in species that have an elongate tube comparable in other ways to the tube of *Chasmanthe*. The difference in the way the floral tube is constructed suggests that adaptation for sunbird pollination may have arisen independently in the two genera. Their stamens also differ in significant ways. In *Chasmanthe* the filaments are slightly unequal in length; the lower abaxial filament, which arches back to lie between the two adaxial filaments, is slightly longer so that the anthers, which are equal in length, reach different levels, the median one protruding beyond the tips of the other two. The anthers are also distinctive in having the lower halves slightly diverging. In *Crocosmia* and other bird-pollinated Iridaceae, the stamens have the same orientation but are equal in length so that the anther tips reach the same point. In *Crocosmia* the anther lobes diverge in the lower third. Lastly, the seeds of *Chasmanthe* are distinctive in the Iridaceae and unique in subfamily Crocoideae, being bright orange, relatively large, and in *C. aethiopica* the seed coat is fleshy and sweet when first exposed on the open capsule. While these features are useful in distinguishing *Chasmanthe* from *Crocosmia*, none of them appears to point toward any relationship with other genera.

Molecular studies conducted at the Jodrell Laboratory at the Royal Botanic Gardens, Kew, in 2002 and continuing have provided surprising new insights about the relationships of genera of Crocoideae. *Chasmanthe* does indeed fall in a clade among genera with excluded seed vasculature, including *Dierama, Duthieastrum, Ixia, Sparaxis* and *Tritonia. Chasmanthe* is not, however, immediately allied with *Tritonia*. Unexpectedly, *Crocosmia* does not fall in this lineage but instead is sister genus to *Devia*, a relationship predicted by Goldblatt and Manning (1990b), and the two genera are together immediately allied to *Freesia*, which does not have excluded seed vasculature. That apparently unusual feature thus seems to have evolved twice within subfamily Crocoideae. *Freesia* does, however, share with *Crocosmia* and *Devia* the specialized leaf anatomical feature, columnar marginal epidermis. An association of *Freesia* with *Crocosmia* at first seemed so unexpected that we reexamined the molecular data, and then when that result was confirmed, we wondered whether the DNA data are in some way not signaling a true relationship. This seems very unlikely, and we have to conclude that evolution in Crocoideae is more complex than until now believed. Features that might be seen to link *Crocosmia, Devia* and *Freesia* include apically notched style branches (more deeply divided in *Freesia*), green, firm-textured floral bracts (later often dry and brown), a capsule with a warty or papillate surface, and a shared chromosome number and karyotype with the basic number $x = 11$.

Further DNA sequence studies will no doubt be undertaken, using different genes or DNA regions, to test the evolutionary relationships of Crocoideae, but we suspect the basic pattern will not change, for it already rests on results of sequences from four DNA regions. We look forward to gaining a deeper understanding of how *Chasmanthe* and *Crocosmia* evolved and how they relate other genera of the subfamily as the DNA of additional species is sequenced, for our current knowledge rests of studies of only one or two examples of each genus.

In the descriptions of the genera and species, measurements are given in metric, and the nonmetric equivalents are approximate.

Chasmanthe

N. E. Brown, Transactions of the Royal Society of South Africa 20: 273 (1932). De Vos, South African Journal of Botany 2: 256 (1985). Type species: *Chasmanthe aethiopica*.

Chasmanthe, from the Greek *chasme*, gaping, and *anthe*, a flower, in allusion to the shape of the flower with the tepals abruptly spreading away from the comparatively wide mouth of the perianth tube.

Deciduous perennials. Rootstock a depressed-globose corm rooting from below, tunics firm-papery, sometimes becoming coarsely fibrous with age. Stem round in cross section, simple or branched. Leaves several, lower two or three forming sheathing cataphylls, foliage leaves unifacial, with a prominent central vein and numerous fine, closely spaced parallel secondary veins, mostly basal and forming a two-ranked fan, the blades sword-shaped, plane, those borne on the stem few and reduced. Inflorescence a spike, flowers either arranged in two ranks (distichous) or on one side of the axis (secund), usually many and crowded; bracts small, green, becoming dry at tips, medium-textured, the inner as long or shorter than the outer and notched apically. Flowers zygomorphic, orange, the lower tepals with contrasting markings (rarely yellow) or the lower laterals entirely green, unscented, with nectar from septal nectaries; perianth tube elongate, cylindric below and sometimes spirally twisted, expanded abruptly and tubular and horizontal above; tepals unequal, the dorsal largest, extended forward and more or less horizontal and concave, remaining tepals much smaller, directed forward or reflexed. Stamens unilateral and arched against the dorsal tepal; filaments arising at base of upper part of tube, the lower (median) one slightly longer than other two; anthers parallel, subversatile, slightly diverging from about the middle. Ovary ovoid, green, concealed by the bracts; style elongate, arching behind the stamens, the branches slightly expanded at the tips. Fruit a globose three-lobed capsule, leathery, sometimes purple on the inside, smooth or lightly warty outside; seeds globose, two to four per locule, orange, shiny and smooth when fresh, the coat sometimes fleshy, then wrinkled on drying, the raphal vascular trace excluded. Basic chromosome number $x = 10$.

Chasmanthe, consisting of three species, is found in the Northern Cape, Western Cape and Eastern Cape provinces of South Africa, extending from coastal central Namaqualand and the Bokkeveld Plateau in the northwest to East London and Kentani in the east, usually in bush, or forest margins.

Chasmanthe has long been thought (for example, Lewis 1954) to be most closely related to *Crocosmia* and *Tritonia,* and Goldblatt and Manning (2000) even expressed doubt about the merit of treating them as separate genera. That assumption, largely based on external appearance, has now been shown by molecular data to be incorrect. Sequences of several chloroplast DNA regions show that *Chasmanthe* belongs in a lineage (or clade) with *Ixia, Sparaxis* and *Tritonia* but surprisingly is most closely related to the southern African genus *Babiana.* Although floral form in *Chasmanthe* and *Crocosmia* is similar, the two genera differ in several technical floral characters, leaf venation and seed features, and they also have different basic chromosome numbers, $x = 10$ in *Chasmanthe* and $x = 11$ in *Crocosmia.* The bright orange seeds of *Chasmanthe,* one species of which has a soft seed coat when first exposed in the capsules, are believed to be adapted for bird dispersal. The flowers are pollinated by sunbirds, and the flowers have the red to orange color, exserted anthers, long perianth tube, extended upper tepal and, reduced lower tepals typical of bird-pollinated flowers in Iridaceae. Similarity to the flowers of the bird-pollinated species of *Crocosmia* is simply the result of convergence for that particular pollination system.

When N. E. Brown described *Chasmanthe* in 1932, he included nine species in the genus, most of them previously placed in the genus *Antholyza,* but also one new species. Two of the nine species of *Chasmanthe* (*C. peglerae* and *C. vittigera*) are now regarded as synonyms of *C. aethiopica* while four others are now known to belong to other genera: *Crocosmia* (*C. fucata*), *Gladiolus* [*Chasmanthe spectabilis* (Schinz) N. E. Brown = *G. magnificus* (Harms) Goldblatt] and *Tritoniopsis* [*C. caffra* (Ker-Gawler ex Baker) N. E. Brown ≡ *T. caffra* (Ker-Gawler ex Baker) Goldblatt and *C. intermedia* (Baker) N. E. Brown ≡ *T. intermedia* (Baker) Goldblatt].

Chasmanthe was not immediately recognized. The South African botanist Edwin P. Phillips (1941) regarded *Chasmanthe* species as belonging in a more broadly defined genus, *Petamenes* Salisbury ex J. W. Loudon, in which Phillips included some 16 species. *Petamenes,* which dates from

1841 but was not recognized by the botanical community at the time, nevertheless has nomenclatural priority over *Chasmanthe,* being an older and validly published name. Careful evaluation of the species once included in *Antholyza* and then removed to various genera, however, including *Petamenes,* has shown that the type species of *Petamenes, P. abbreviatus* (Andrews) N. E. Brown, belongs in *Gladiolus.* Thus *Petamenes* becomes a nomenclatural synonym of *Gladiolus* (Goldblatt and Manning 1998), and *Chasmanthe* is now recognized in a somewhat emended circumscription, with some erstwhile *Petamenes* species transferred to *Crocosmia* as well as to *Gladiolus* and *Chasmanthe.*

Key to *Chasmanthe*

1 Stem curved at the base of the inclined to horizontal spike and flowers carried on the upper side of the axis, usually in one rank; slender part of the perianth tube spirally twisted; seeds fleshy when first exposed, later becoming dry and wrinkled, the inside of the capsule reddish to purple . . . . *C. aethiopica*
1 Stem and spike erect, often branched and flowers carried in two ranks; slender part of the perianth tube not spirally twisted; seeds dry, shiny, with a hard seed coat, the inside of the capsule straw-colored 2
2 Tepals orange to scarlet, the lower more or less spreading, except the lower median; capsule warty and with a nipple-like apex; the upper tepal arising $^{1}/_{8}$–$^{1}/_{4}$ inch (3–7 mm) beyond the lower tepals . . . . *C. floribunda*
2 Upper and lower medial tepals orange to scarlet but the lower laterals green, all directed forward; capsule smooth, rounded at the apex; all the tepals arising at about the same level . . . . *C. bicolor*

Chasmanthe aethiopica (Linnaeus) N. E. Brown

PLATE 1

N. E. Brown, Transactions of the Royal Society of South Africa 20: 273 (1932). De Vos, South African Journal of Botany 51: 256 (1985).

Antholyza aethiopica Linnaeus, Systema Naturae, ed. 10, 2: 863 (1759). *Petamenes aethiopica* (Linnaeus) Phillips, Bothalia 7: 44 (1941). Type: South Africa, without precise locality or collector (Herb. Linnaeus 60.3, LINN, holotype).
Antholyza vittigera Salisbury, Transactions of the Horticultural Society of London 1: 324 (1812). *Chasmanthe vittigera* (Salisbury) N. E.

Brown, Transactions of the Royal Society of South Africa 20: 274
(1932). *Petamenes vittigera* (Salisbury) Phillips, Bothalia 7: 44
(1941). Type: South Africa, Western Cape, without precise locality
or collector, illustration in Curtis's Botanical Magazine 29: plate
1172 (1809), lectotype designated by de Vos (1984).

Antholyza emarginata Thunberg ex Baker, Handbook of the Irideae,
page 230 (1892). Type: South Africa, Western Cape, without precise
locality, *C. P. Thunberg s.n.* (Herb. Thunberg 1100, UPS, holotype).

Chasmanthe peglerae N. E. Brown, Transactions of the Royal Society of
South Africa 20: 273 (1932). *Petamenes peglerae* (N. E. Brown)
Phillips, Bothalia 7: 44 (1941). Type: South Africa, Eastern Cape,
Kentani division, near Black Rock Cove, July 1902, *A. M. Pegler 500*
(K, holotype; BOL, PRE, isotypes).

aethiopica, a classical Latin adjective for a plant from Africa south of
the Sahara.

Plants 16–26 inches (40–65 cm) high. Corm depressed-globose, $1^3/_{16}$–
$1^9/_{16}$ inches (30–45 mm) in diameter, with tunics papery to fibrous. Stem
unbranched. Leaves sword-shaped, medium-textured, with a prominent
midrib, mostly $^1/_2$–$^{13}/_{16}$ inch (12–20 mm) wide. Spike arching outward to
almost horizontal, 10- to 16-flowered, the flowers borne on the upper
side; bracts green or flushed red on the margins or throughout, $^5/_{16}$–$^9/_{16}$

Chasmanthe aethiopica

inch (8–12(–15) mm) long, the inner minutely forked at the tip. Flowers facing to the side or toward the base of the spike, bright orange to scarlet, the tube often yellowish on the underside, unscented; perianth tube trumpet-shaped, slender below for $^3/_{16}$–$^9/_{16}$ inch ((5–)7–10(–15) mm) and twisted through 360°, abruptly expanded into a wide, cylindric upper part $^5/_8$–1 inch (16–25) mm long, lightly three-pouched at the base, arching outward; tepals unequal, the dorsal largest, $^7/_8$–1$^3/_8$ inches × $^1/_4$–$^3/_8$ inch ((22–)25–35 × 7–10 mm), spathulate and spooned, the upper and lower lateral tepals mostly $^3/_8$–$^9/_{16}$ inch (10–15 mm) long, spreading or half-reflexed, the lowermost smallest, channeled, directed more or less forward. Filaments arching under the dorsal tepal, 1$^9/_{16}$–2 inches (40–50 mm) long, the median one $^1/_8$ inch (3–4 mm) longer; anthers $^3/_{16}$–$^1/_4$ inch (5–7 mm) long, the median reaching almost to the dorsal tepal apex, light purple, the pollen purple. Style arching behind the filaments, dividing opposite the upper third of the anthers, often ultimately exceeding them, the branches $^1/_8$–$^3/_{16}$ inch (4–5 mm) long. Capsules ovoid to barrel-shaped and three-lobed, smooth, dark red to purple-black inside on opening; seeds as many as three per locule, more often one or two (that is, three to nine per capsule), globose, initially $^1/_4$ inch (6–7 mm) in diameter, bright orange, the coat fleshy and smooth, becoming wrinkled on drying and $^3/_{16}$–$^1/_4$ (5–6 mm) in diameter. Chromosome number $2n = 20$. Flowering time mid-autumn to early winter (April–June), sometimes in midwinter (July).

Occurring mainly in the winter-rainfall zone of southern Africa, the range of *Chasmanthe aethiopica* extends from Darling, 37 miles (60 km) north of Cape Town, in Western Cape province in the west, along South Africa's southern coast to Kei Mouth and Kentani in Eastern Cape province in the east. Plants grow in coastal bush and along forest margins in a variety of soils.

The long-tubed flowers are adapted for pollination by sunbirds, of which the lesser double-collared sunbird, *Nectarinia afra,* is the most frequently recorded visitor. This sunbird favors bushy coastal habitats and is thus the most likely pollinator. A second sunbird, *N. famosa,* the malachite sunbird, has also been reported as a visitor to *Chasmanthe aethiopica* (see the chapter, Ecology, under Pollination). The large, round, orange seeds have a fleshy coat when first exposed, suggesting that they not only attract birds to feed on them but offer a certain amount of nutrition as well. The seed coat has a slightly sweet taste and a leathery texture.

Chasmanthe aethiopica is readily recognized in the genus by its comparatively short size, the spike seldom reaching more than 20 inches (50 cm) in height, the consistently unbranched flowering stem and the inflexed and inclined to almost horizontal spike. In addition, the 10–16 flowers are borne in a single congested row on the upper side of the spike axis. The flowers, too, are distinctive for the perianth tube is abruptly expanded above the narrow cylindric lower part, the wider upper part of the tube is lightly expanded into three pouches, and the lower part of the tube is twisted through 360°. As in other species of *Chasmanthe,* the tepals are unequal, with the dorsal largest, the smaller upper lateral, and lower lateral tepals spreading to lightly reflexed, and the small, narrow, lowermost tepal channeled and directed forward. While all *Chasmanthe* species have bright orange seeds, those of *C. aethiopica* are the only ones that have a fleshy seed coat when the they are first exposed after the capsule walls split. The seed coat later becomes dry and wrinkled. The inside of the capsules is often flushed reddish to purple. Thus both vegetatively, and in flower and fruit, *C. aethiopica* stands apart from its two congeners.

So much confusion surrounds the identity and correct name for *Chasmanthe aethiopica* that we can only shake our heads in dismay. The name *C. aethiopica* is a combination based on *Antholyza aethiopica,* described in 1759 by Carl Linnaeus, who based the species both on preserved herbarium specimens and old woodcut illustrations. The earliest and best of the latter is the figure published in a 1635 volume by Jacques-Philippe Cornut, *Canadensium Plantarum.* Named in that volume, *Gladiolus aethiopicus flore coccineo* (red-flowered *Gladiolus* from Africa), the woodcut unmistakably represents what we now know as *C. floribunda.* So similar, however, were the flowers of the plant represented in Cornut's work (and in later copies published elsewhere) to dry plant specimens of *C. aethiopica* that Linnaeus had acquired, perhaps directly from South Africa, that he considered them the same species. Thus for more than 50 years, *C. aethiopica* and *C. floribunda* were not perceived as different in any significant way. In 1802 the British bulb expert John Ker-Gawler identified the painting of *C. floribunda* that appeared in plate 561 of *Curtis's*

Botanical Magazine as *Antholyza aethiopica*. Then, in 1809, Ker-Gawler, who should have known better, identified the painting of the real *C. aethiopica* represented in plate 1172 of *Curtis's Botanical Magazine* as 'A. aethiopica* var. β'. At least he did recognize that this plant differed from the one he mistakenly thought was *A. aethiopica*, even though he did not appreciate the significance of those differences.

It was only after the gifted but idiosyncratic English gardener and botanist Richard A. Salisbury saw both *Chasmanthe floribunda* and *C. aethiopica* in bloom that it became clear, at least to him, that there were two species called *Antholyza aethiopica*. In 1812 Salisbury described *A. floribunda*, the type of which is now fixed by convention to be the 1802 illustration on plate 561 in *Curtis's Botanical Magazine*. Salisbury incidentally gave the name *A. vittigera* to the plants we now know as *C. aethiopica*, conveniently avoiding use of the older Linnaean name *A. aethiopica*. By sheer coincidence, the great French flower painter and an acute observer Pierre Redouté published a painting of *A. floribunda* in the sumptuous *Les Liliacées* in 1813, calling it *A. prealta*. Evidently both Salisbury and Redouté independently reached the same conclusion about the taxonomy of plants till then called *C. aethiopica*. Their decision was not always accepted, for J. G. Baker, working at the end of the 19th century, regarded the two species merely as varieties of one another. In 1917 the botanist Rudolf Marloth, working in South Africa, recognized both species, as *A. aethiopica* and *A. prealta*. N. E. Brown was equally convinced, and when he described the genus *Chasmanthe* in 1932, he recognized both *C. aethiopica* and *C. floribunda*, with *A. prealta* a synonym of the latter. E. P. Phillips likewise accepted the two species but preferred to use the name *Petamenes* for all *Chasmanthe* species.

Lack of understanding of *Chasmanthe aethiopica* and its pattern of variation led N. E. Brown to recognize another species, *C. peglerae*, based on plants from the eastern part of the range of *C. aethiopica*, in the Kentani district of Eastern Cape province, as well as Salisbury's *C. vittigera*. These proved on more careful examination to be nothing but the most trivial variants of *C. aethiopica* and were not recognized taxonomically, even at varietal rank, by M. P. de Vos in her 1985 account of the genus *Chasmanthe*.

Chasmanthe bicolor (Gasparrini) N. E. Brown
PLATE 2

N. E. Brown, Transactions of the Royal Society of South Africa 20: 270 (1932). De Vos, South African Journal of Botany 51: 259 (1985).

Antholyza bicolor Gasparrini, Annali Civili Regno delle Due Sicilie 1(4): 118 (1832). *Antholyza aethiopica* var. *bicolor* (Gasparrini) Baker, Journal of the Linnean Society 16: 179 (1877b). *Petamenes bicolor* (Gasparrini) Phillips, Bothalia 7: 44 (1941). Type: South Africa, Western Cape, without precise locality or collector, illustration in C. J. E. Morren, La Belgique Horticole 2: plate 25/1 (1852), lectotype designated by de Vos (1985).

Antholyza aethiopica var. *minor* Lindley, The Botanical Register 14: plate 1159 (1828). Type: South Africa, without precise locality or collector, illustration in The Botanical Register 14: plate 1159 (1828), lectotype designated by de Vos (1985).

bicolor, Latin for two-colored, alluding to the predominantly orange-scarlet perianth with green markings.

Plants 28–36 inches (70–90 cm) high. Corm lightly depressed-globose, $^{13}/_{16}$–1 inch (20–25 mm) in diameter, the tunics thin-papery. Stem erect, usually with one to three short branches, the spike strongly drooping in bud and flowering early. Leaves sword-shaped, with a prominent central vein, reaching to about the base of the spike, $^{11}/_{16}$–1$^5/_{16}$ inches (17–33 mm) wide. Spike erect, with flowers in two opposed ranks, mostly 18- to 28-flowered; bracts green, with dry tips, $^1/_4$–$^3/_8$ inch (6–10 mm) long. Flowers secund, orange-scarlet, the lower lateral tepals green and the tube yellow inside and on the outside below, unscented; perianth tube trumpet-shaped, slender below $^1/_4$–$^3/_8$ inch (6–10 mm), abruptly expanded into a wide and cylindric upper part mostly $^{13}/_{16}$–1 inch (20–25 mm) long, arching outward and horizontal above; tepals all arising at about the same level, unequal, the dorsal largest, 1$^1/_{16}$–1$^9/_{16}$ inches (27–40 mm) long, spathulate and folded along the midline, upper lateral and lowermost tepals directed forward, $^3/_{16}$–$^5/_{16}$ inch (5–8 mm) long, the lower laterals spreading to half-recurved, about $^5/_{16}$ inch (8 mm) long. Filaments arching under the dorsal tepal, 1$^3/_4$–2$^3/_8$ inches (45–60 mm) long; anthers $^1/_4$ inch (6–7 mm) long, the median usually exceeding the dorsal tepal, purple, the pollen purple. Style arching behind the filaments, dividing opposite the middle of the anthers, the branches about $^1/_8$ inch (3 mm) long. Capsules more or less ovoid and three-lobed, showing the outline of the seeds, $^9/_{16}$–$^{13}/_{16}$ inch (15–20 mm) long, smooth; seeds globose, two per locule, $^1/_8$ inch (3.5–4 mm) long, yellowish to orange, the coat hard, semimatte. Chromosome number unknown. Flowering time midwinter to early spring (July–August).

A narrow endemic of Western Cape province, South Africa, the native geographic range of *Chasmanthe bicolor* is not well known, surprising for an area that has been studied botanically for more than two centuries. Although it was known in gardens in Europe before 1828, precisely where *C. bicolor* occurs in the wild has long been puzzling. There are only two reliable records. One is the Vrolikheid Nature Reserve near MacGregor, in the Robertson district of Western Cape province. There, a collection from an apparently wild population was made in 1988. The second, made in 2003, is from near the large southern Cape town of Swellendam, a short distance east of MacGregor. Other records are, we believe, all from cultivated plants, and the early 20th century record from the well-collected

Chasmanthe bicolor

Franschhoek Valley has never been confirmed. In the faint hope of finding a wild population, we (J.M. and P.G.) undertook as search for *C. bicolor* in early spring 2002 at the Vrolikheid Nature Reserve. This delightful and botanically fascinating area, with well-preserved native flora, and
wild springbok and ostrich, seemed to have ample suitable habitat for *C.
bicolor,* but we found no plants of the species there. It may have been a
little too late to find plants in flower, and locating them in fruit in the relatively large area of the reserve is difficult if not impossible. Then quite by
chance in August 2003, we (J.M.) encountered a small colony of wild *C.
bicolor* between Swellendam and the village of Suurbraak at the foot of
the Langeberg. There, plants grew along a stream in the shade of evergreen riverine woodland that included *Buddleja saligna, Grewia occiden-
talis, Olea africana* and *Sideroxylon inerme.* The tantalizing botanical mystery about the native habitat of *C. bicolor* now seems to have been solved.

Despite the historical confusion about the native range of *Chasmanthe
bicolor,* there can be no doubt that it is a bona fide species. The almost
globose corm, unlike the disk-like corm of its two congeners, the slender
shape of the flower with small lateral and lower tepals, the green color of
the lower lateral tepals, the yellow underside of the tube and the short
style branches, about $1/8$ inch (3 mm) long, are diagnostic for the species.
The strongly drooping apex of the spike, evident even after the lower
flowers have opened, is also distinctive. The spike usually becomes erect
before the first flower buds open in *C. aethiopica* and *C. floribunda. Chasmanthe bicolor* appears to be most closely related to the more robust *C.
floribunda,* which has a particularly large corm, larger, bright orange-scarlet flowers, and style branches $1/4–3/8$ inch (7–10 mm) long. *Chasmanthe
floribunda* also has distinctive capsules with a nipple-like apex, lacking in
C. bicolor.

Free-flowering, *Chasmanthe bicolor* is an attractive horticultural subject. It lacks the vigor of *C. floribunda* and the flowers are slightly smaller
and less brilliantly colored, but they nevertheless have an interesting form,
and the combination of green and yellow in a predominantly orange-
scarlet bloom is quite striking.

We have no record of how the rare *Chasmanthe bicolor* came into cultivation. It first appeared in the botanical literature in 1828 when it was
described by the eminent English botanist John Lindley in an illustrated

article in *The Botanical Register*. Lindley regarded *C. bicolor* as merely a variety of the well-known *C. aethiopica,* then called *Antholyza aethiopica,* which at the time was not recognized as separate from *C. floribunda*. According to Lindley, the plant had been in cultivation in European gardens for many years. Specimens preserved in the herbarium of the natural history museum in Vienna bearing the date 1811 attest to Lindley's assertion.

Named independently at species rank, *Antholyza bicolor* was described in 1832 by Guglielmo Gasparrini in an account of the plants cultivated at the Royal Botanic Garden at Boccadifalco, near Palermo in Sicily. In a second article dealing with plants grown at the Royal Botanic Garden, Naples, by Michele Tenore (1845), Gasparrini is again credited with the authorship of the species, confirming that it was being grown in Naples. That *Chasmanthe bicolor* was already in cultivation in Sicily in 1832 and in Naples in 1845 lends additional credence to Lindley's statement that the species had long been grown in gardens.

Chasmanthe floribunda (Salisbury) N. E. Brown

PLATE 3

N. E. Brown, Transactions of the Royal Society of South Africa 20: 274 (1932). De Vos, South African Journal of Botany 51: 258 (1985).

Antholyza floribunda Salisbury, Transactions of the Horticultural Society, London 1: 324 (1812). *Petamenes floribunda* (Salisbury) Phillips, Bothalia 7: 44 (1941). Type: South Africa, without precise locality or collector, illustration in Curtis's Botanical Magazine 16: plate 561 (1802), lectotype designated by de Vos (1985).
Antholyza prealta Redouté, Les Liliacées 7: plate 387 (1813). Type: South Africa, without precise locality or collector, illustration in Redouté, Les Liliacées 7: plate 387 (1813).
Chasmanthe floribunda var. *duckittii* G. J. Lewis in H. M. L. Bolus, South African Gardening 23: 47 (1933). De Vos, South African Journal of Botany 51: 259 (1985). Type: South Africa, Western Cape, near Darling, cultivated at Kirstenbosch, *F. Duckitt s.n.* (National Botanical Garden 419/27, BOL, holotype).

floribunda, Latin for 'producing many flowers', referring to the many-flowered spikes.

Plants 20–47 inches (50–120 cm) high. Corm depressed-globose, $2^3/_8$–$2^3/_4$ inches (6–7 cm) in diameter when mature, with tunics firm-papery, becoming fibrous with age. Stem erect, usually branched, the branches ascending. Leaves sword-shaped, medium-textured, with a prominent midrib, mostly 1–$1^3/_8$ inches (25–35 mm) wide, occasionally barred light purple on the sheaths. Spike erect, with flowers in two opposed ranks, mostly 30- to 40-flowered; bracts $^3/_8$–$^9/_{16}$ inch ((9–)13–15 mm) long, green, flushed red on the margins or throughout. Flowers deep orange to scarlet (rarely yellow), the lower half of the tube paler orange or yellow, unscented; perianth tube trumpet-shaped, slender below for $^3/_{16}$–$^5/_{16}$ inch ((5–)7–8 mm) and spirally twisted, abruptly expanded into a wide and cylindric upper part $^{13}/_{16}$–1 inch (20–25 mm) long, slightly pouched at the base, arching outward and horizontal above; tepals unequal, the dorsal largest, $^{13}/_{16}$–$^{15}/_{16} \times ^1/_4$–$^5/_{16}$ inch ((21–)24–26 × 6.5–8 mm), spathulate and spooned, the upper lateral tepals about $^1/_2$ inch (12 mm) long, the lower laterals about $^9/_{16} \times ^1/_4$ inch (15 × 6 mm), the lower median about $^3/_8$ inch (10 mm) long. Filaments arching under the dorsal tepal, the median

Chasmanthe floribunda

$^1/_8$–$^1/_4$ inch (4–6 mm) longer; anthers $^1/_4$ inch (6–7 mm) long, parallel, the median extending beyond the dorsal tepal apex, purple, the pollen purple. Style arching over the filaments, usually dividing opposite the middle of the lateral anthers, sometimes at the base of the anthers, the branches $^1/_4$–$^3/_8$ inch (7–10 mm) long. Capsules depressed-globose and trilobed, showing the outline of the seeds, $^3/_8$–$^9/_{16}$ inch (10–15 mm) long, tapering to a nipple-like point, the surface lightly warty; seeds as many as 4 per locule, thus as many as 12 per capsule, globose or somewhat angled by pressure, about $^3/_{16}$ inch (5 mm) in diameter with a flattened or sunken chalazal end, bright orange, the coat hard and shiny. Chromosome number $2n = 20$. Flowering time midwinter to spring (July–September).

Occurring in the winter-rainfall zone of South Africa, *Chasmanthe floribunda* extends from coastal Namaqualand in the north to Hermanus, Western Cape province, in the south. Plants are most common along the coast and in the coastal mountains and favor sandstone and granite substrates. Somewhat surprisingly, the species also occurs inland in the western karoo near Nieuwoudtville, Northern Cape province, and in the nearby Kobee Mountains. In the western karoo, however, plants grow on doleritic clay around large dolerite boulders, and in the Kobee Mountains on cool, south-facing slopes along streams in shale.

Like all species of *Chasmanthe,* the flowers of *C. floribunda* are adapted for pollination by sunbirds. The malachite sunbird, *Nectarinia famosa,* is the most frequent visitor in our experience, but the lesser double-collared sunbird, *N. chalybea,* may also be seen visiting the flowers. So attractive and rewarding are the flowers that it is seldom that sunbirds are not seen in the vicinity of flowering individuals, and particularly in the early morning or later afternoon, if one watches quietly and out of sight, it is rare not to see a sunbird fly to a flowering spike, alight and begin probing flowers for their generous supply of nectar.

A strikingly attractive plant when in flower, *Chasmanthe floribunda* makes an elegant and desirable garden subject. The tall, branching stems, bearing numerous brightly colored, large flowers, assure at least 4 weeks of bloom, while the sword-shaped leaves, arranged in erect, two-ranked fans, constitute a striking accent for several months. Even the fruiting spikes, which tend to turn reddish with age except in the yellow-flowered form, remain attractive. Like other *Chasmanthe* species, *C. floribunda* is deciduous, dying back in the summer, when it is best left in the ground dry, or the corms may be lifted and stored in a well-ventilated place till autumn.

The species is readily distinguished from its congeners by the relatively large flowers, branched stem, erect spike and uniformly colored, bright orange-scarlet tepals borne on an erect, two-ranked spike. Long confused with *Chasmanthe aethiopica,* especially in the 18th and 19th centuries, that species is shorter and has an unbranched flowering stem with an inflexed spike, either strongly inclined or nearly horizontal, with the flowers borne in a single row on the upper surface of the flowering axis. A curious feature of *C. aethiopica* is the twisted lower, narrow part of the perianth tube, which makes the flowers of that species unmistakable.

Although relatively uniform over most of its range, the inland populations of *Chasmanthe floribunda,* from the western karoo near Nieuwoudtville, occur in an unusual habitat: rocky dolerite ground in bush at the base of huge dolerite boulders. Undoubtedly *C. floribunda,* they nevertheless have some unusual features, most obviously the purple banding on the leaf sheaths and the paler orange flowers. The flowers are also slightly smaller than the well-known coastal form in some respects, notably the shorter perianth tube, as long as 1 inch (25 mm), the dorsal tepal about $^{13}/_{16}$ inch (21 mm) long and the style dividing below the bases of the anthers, when more typically it divides opposite the middle of the anthers. These relatively small differences are consistent with an outlying race of the species growing on a different substrate. Elsewhere, *C. floribunda* is typically found on sandstone or granite substrates.

A color variant, sometimes called variety *duckittii,* now widely cultivated in South Africa, was descended from a single plant or small colony found in the 1920s near Darling, north of Cape Town. The flowers are an attractive pale yellow, and the anthers and pollen brownish. We do not regard the plant as worth taxonomic recognition. It is no more than a color sport or spontaneous mutant. Horticultural recognition as a cultivar seems appropriate. Both color variants of *Chasmanthe floribunda* are excellent plants for the garden in areas of mild winters, especially those with Mediterranean climates such as California and Western Australia.

They also thrive in the somewhat frosty, dry winters of the South African high veld where they must be watered during the growing season. Not only do the plants make a striking display for 3–4 weeks in spring, but the upright elegant foliage is attractive for another month while the fruits develop. Of course, plants die down in summer, even in places where there is ample summer rain. They can be interplanted with a summer-blooming bulb so the ground is not left bare while they are dormant. The deciduous, pale-blue-flowered *Agapanthus campanulatus* is an ideal companion to *C. floribunda* in the garden.

Chasmanthe floribunda, well named for its profusion of flowers, has a long and muddled history, for until 1812 it was confused with its less-attractive sister species, *C. aethiopica*. Their history is inextricably intertwined and is laid out in more detail in the discussion of *C. aethiopica*. Linnaeus cited examples of both species in the protologue of *Antholyza aethiopica,* and later experts, including Linnaeus's successor, Carl Peter Thunberg, the Austrian and German experts Johann Jacob Roemer and Joseph August Schultes (1817) and even John Ker-Gawler and John Gilbert Baker, regarded them as the same species. Ker-Gawler was evidently the first to at least see some difference between them, for he named an 1809 painting (of *C. aethiopica*) *A. aethiopica* var. β, comparing it to the 1802 painting of *C. floribunda* that he called *A. aethiopica*.

Then, Richard Salisbury in 1812 and Pierre Redouté in 1813, each with knowledge of living plants of *Chasmanthe aethiopica* and *C. floribunda,* published new names for the latter, of course placing the plants in the genus *Antholyza*. Salisbury's name, *A. floribunda,* being earlier, is the one we use today, but Redouté's *A. prealta* was accepted by some later botanists in disregard for the rules of priority in plant nomenclature. Thus the Friedrich Klatt used the name *A. prealta* in 1882, as did Rudolf Marloth in 1917, for what Nicholas Brown recognized as *C. floribunda* in 1932, when he separated *Chasmanthe* species from *Antholyza* in the course of dismantling that genus.

Crocosmia

Planchon, Flore des Serres et des Jardins de l'Europe 7: 161 and plate 702 (1851). De Vos, Journal of South African Botany 50: 474 (1984). Type species: *Crocosmia aurea.*

Tritonia section *Crocosmia* (Planchon) Baker, Journal of the Linnean Society 16: 163 (1877b), as '*Crocosma*'.
Crocanthus Klotzsch in manuscript, mentioned in synonymy by Klatt in Peters, Naturwissenschaftliche Reise nach Mossambique, Botanik 2: 516 (1864), without description and an invalid name.
Curtonus N. E. Brown, Transactions of the Royal Society of South Africa 20: 270 (1932). Type: *C. paniculatus* (Klatt) N. E. Brown (≡ *Crocosmia paniculata*).

Crocosmia, from the Greek *krokos,* that is, crocus, the source of the culinary herb saffron, and *osme,* scent, thus 'smelling of saffron'; the grammatically more correct form *Crocosma* was often used in the 19th century but *Crocosmia* now seems firmly established. The flowers also produce an orange, saffron-colored dye.

Deciduous perennials. Rootstock a depressed-globose corm rooting from below, those of past seasons often not resorbed, producing thin rhizomes from the base in some species, tunics firm-papery, sometimes becoming fibrous with age. Stem round in cross section, rarely ridged, simple or branched, sometimes repeatedly and strongly flexuose. Leaves several, the lower two or three forming sheathing cataphylls, foliage leaves unifacial, either plane and medium-textured with a prominent central vein, or pleated and firm-textured with a prominent vein at each fold, mostly basal and forming a two-ranked fan, the blades lanceolate to sword-shaped, those borne on the stem few and reduced. Inflorescence a spike, flowers either arranged in two ranks (distichous) or lying on one side of the axis (secund), usually many and crowded, the axes often strongly flexuose; bracts small, green, becoming dry at the tips, firm-textured, the inner about as long as the outer and notched apically. Flowers zygomorphic and funnel-shaped to tubular or actinomorphic and nodding, in shades of yellow or orange to scarlet, lower tepals sometimes with contrasting markings or green, unscented, with nectar from septal nectaries; perianth tube cylindric or funnel-shaped, widening gradually and flared or tubular above; tepals equal or unequal, then the dorsal largest. Stamens unilateral and arched, sometimes lying against the dorsal tepal, or symmetrically disposed and central; filaments arising at base of the upper part of the tube, included to well exserted; anthers parallel, subversatile, slightly diverging in the lower third. Ovary ovoid, green, enclosed by the bracts; style exserted, ascending or arching behind the horizontal stamens, the branches filiform or expanded apically, notched at the tips. Fruit a depressed-globose three-lobed capsule, leathery, often warty above, sometimes reddish to orange on inside; seeds globose or angular and prismatic, several to only two per locule, brown or reddish to black, either shiny and smooth when fresh, the coat at least sometimes fleshy and wrinkled on drying, or soft and spongy, the raphal vascular trace excluded. Basic chromosome number $x = 11$.

Crocosmia, consisting of eight species, is confined to Africa south of the Sahara, and Madagascar. Species are concentrated in eastern southern Africa, including Lesotho, South Africa, Swaziland and Mozambique, where six occur. One is endemic to Namaqualand, Northern Cape province, South Africa, and one is restricted to western Madagascar (Goldblatt and Manning 1990a). The most widespread species, *C. aurea,* extends from Eastern Cape to Angola in the west and Uganda and Central African Republic in the north. Plants occur in grassland, often in rocky sites, or forest margins, but the Namaqualand species, *C. fucata,* grows in shade in riverine scrub. *Crocosmia aurea* is often found in the deep shade of the floor of evergreen forest.

Relatively diverse florally, *Crocosmia* species show a fair degree of floral adaptations. The short-tubed *C. mathewsiana* and *C. pottsii* are pollinated by large anthophorine bees and *Apis mellifera* whereas the actinomorphic-flowered *C. aurea* is probably pollinated by large butterflies. Species with elongate floral tubes are pollinated by sunbirds (*Nectarinia*).

Crocosmia closely resembles the larger southern and tropical African genus *Tritonia,* which has some 28 species with orange flowers and short floral bracts. That resemblance is, however, misleading. DNA sequence information from several chloroplast genes or DNA regions shows that *Crocosmia* is most closely related to *Devia,* a local endemic with a single species of the Roggeveld Plateau near Sutherland in South Africa, and together the two genera are related to the African genus *Freesia*, with 16 species. Similarities between *Crocosmia* and *Chasmanthe* in flower form and general habit, long thought to indicate close relationship, are likewise misleading. *Chasmanthe* is related to *Tritonia*, though not its immediate ally. Similarities shared between *Crocosmia* and *Chasmanthe* are probably associated with their pollination systems. Flowers of three species of *Crocosmia* and all three of those of *Chasmanthe* are adapted for pollination by sunbirds (*Nectarinia*), and like all bird-pollinated flowers they have reddish to orange pigmentation, long stamens and an elongate, relatively wide perianth tube.

Species of *Tritonia* can be distinguished from those of *Crocosmia* by their softer-walled capsules with several small, hard seeds per locule, often one or more of the lower tepals with a prominent tooth-like callus and floral bracts that are dry and membranous rather than leathery as in *Crocosmia*.

Crocosmia species described before 1850 were referred either to *Tritonia* or, when long-tubed, to *Antholyza*. *Crocosmia* was described in 1851 by Jules Émile Planchon, for the distinctive *C. aurea*, which has nodding, radially symmetric flowers with elongate, well-exserted stamens. Curiously, we know from an unpublished name, listed in an account of the plants collected by the German explorer Wilhelm Peters, that his compatriot Johann Friedrich Klotzsch intended to place *C. aurea*, which he knew from Peters's collection from Mozambique, in a new genus at about the same time that Planchon named *Crocosmia*. Klotzsch's unpublished name was *Crocanthus*—we wonder whether the similarity to Planchon's choice was coincidence, influenced by the same saffron-like odor, or was *C. aurea* being grown in gardens in Europe and known by such names prior to the formal publication of the name *Crocosmia*?

Subsequent to 1851, short-tubed species now included in *Crocosmia* were placed in *Tritonia* or *Montbretia*, a genus described in 1811 by Augustin Pyramus de Candolle and now a nomenclatural synonym of *Tritonia* (the type species of which is *T. laxiflora*). *Crocosmia* was finally accepted by J. G. Baker in 1892 and by F. W. Klatt in 1894, both of whom spelled the name *Crocosma*.

The English botanist N. E. Brown saw the similarity between species of *Tritonia*, especially those included in section *Montbretia*, and *Crocos-*

mia aurea, and in 1932 he transferred *T. masoniorum* and *T. pottsii,* as well as two more that we do not recognize today at species rank, to *Crocosmia.* One long-tubed species, then *Antholyza fucata,* was included in *Chasmanthe* by Brown, while *A. paniculata,* which has long-tubed flowers like those of *C. fucata* but pleated leaves, was placed by Brown in a new genus, *Curtonus.* The obvious similarity of the plant placed in *Curtonus* to species of *Crocosmia,* in the pleated leaves, found in *Crocosmia masoniorum* and *C. masoniorum* (and *C. mathewsiana,* which Brown left in *Tritonia*) and in the short floral bracts and other critical floral features, persuaded one of us (Goldblatt 1971) that *Curtonus* should be included in *Crocosmia.* That decision was supported by similarities in the chromosome cytology of *Curtonus,* which has small chromosomes and a basic number x = 11, matching exactly the situation in *Crocosmia.*

Crocosmia fucata and *C. mathewsiana* were transferred to the genus by Miriam de Vos in 1984 when she revised *Crocosmia.* The yellow-flowered Madagascan plant described as *Geissorhiza ambongensis* in 1939 was transferred to *Crocosmia* by Goldblatt and Manning in 1990.

Key to *Crocosmia*

1 Leaves sword-shaped and plane, with a prominent central vein 2
1 Leaves lanceolate and pleated, with several veins more or less equally
  prominent 6
2 Flowers actinomorphic and nodding, tepals all spreading from the base;
  filaments well exserted from the tube and directed downward . . . . *C. aurea*
2 Flowers zygomorphic and secund or half-nodding, the dorsal tepal
  somewhat to much larger and the stamens unilateral; filaments largely
  exserted from the tube, suberect or horizontal 3
3 Perianth tube elongate and wide, and cylindric in the upper half; the
  dorsal tepal extended more or less horizontally . . . . *C. fucata*
3 Perianth tube funnel-shaped, the upper part flaring and widest at the
  mouth 4
4 Plants as tall as 4 inches (10 cm), usually unbranched; spikes with 1–3
  flowers (plants of Madagascar) . . . . *C. ambongensis*
4 Plants 20–40 inches (50–100 cm) tall, usually with several branches;
  spikes usually with more than 15 flowers 5
5 Filaments $^5/_{16}$–$^1/_2$ inch (8–12 mm) long; flowers funnel-shaped with
  the tepals directed forward (plants of southern Africa) . . . . *C. pottsii*

5 Filaments $^9/_{16}$–1$^9/_{16}$ inches (15–40 mm) long; flowers salver-shaped with
   the tepals spreading in the upper half (see Hybrids) .... *C.* ×*crocosmiiflora*

6 Perianth tube $^9/_{16}$–1 inch (15–25 mm) long, the upper part funnel-
   shaped and flaring; filaments well exserted from the flower, 1$^3/_{16}$–1$^3/_8$
   inches (30–35 mm) long 7

6 Perianth tube 1$^5/_{16}$–1$^3/_4$ inches (33–45 mm) long, the upper part wide and
   cylindric or weakly flaring; filaments $^1/_2$–1$^3/_8$ inches (13–35 mm) long ... 9

7 Flowering stem usually unbranched; spike arching outward to nearly
   horizontal; dorsal tepal $^{13}/_{16}$–1$^3/_{16}$ inches (20–30 mm) long
       .... *C. masoniorum*

7 Flowering stem somewhat to much branched; dorsal tepal $^9/_{16}$–$^7/_8$ inch
   (14–22 mm) long 8

8 Flowering stem branched repeatedly; spikes drooping; dorsal tepal
   about $^9/_{16}$ inch (14 mm) long; filaments $^1/_2$–$^9/_{16}$ inch (13–15 mm) long
       .... *C. mathewsiana*

8 Flowering stem with just a few branches; spikes arching outward;
   dorsal tepal $^7/_8$ inch (22 mm) long; filaments $^{11}/_{16}$–$^{13}/_{16}$ inch (17–20
   mm) long .... *C.* ×*crocosmoides*

9 Dorsal tepal $^9/_{16}$–$^{11}/_{16}$ inch (14–18 mm) long; plants often growing in
   clumps; flowering stem divaricately branched and the branches often
   also branched; spike axes strongly flexuose .... *C. paniculata*

9 Dorsal tepal $^{13}/_{16}$–1$^3/_{16}$ inches (20–30 mm) long; plants usually solitary;
   flowering stem simple or few-branched; spike axes hardly flexuose or
   straight .... *C. pearsei*

Crocosmia ambongensis (H. Perrier) Goldblatt and J. C. Manning
PLATE 4

Goldblatt and J. C. Manning, Bulletin du Museum National d'Histoire
Naturelle, série 4, section B, Adansonia 12: 59–64 (1990a).

Geissorhiza ambongensis H. Perrier, Notulae Systematicae 8: 130–131
(1939). Type: Madagascar, Majunga, Tsingy de Namoroka, near An-
dranomavo, February 1903, *H. Perrier 1519* (P, lectotype, designated
by Goldblatt and Manning 1990a).

ambongensis, Latin for 'from Ambong' a region in Mahajunga province,
Madagascar, where the species was first found.

Plants 2$^3/_8$–4 inches (6–10 cm) high. Corm globose, $^1/_2$–$^9/_{16}$ inch (12–15
mm) in diameter, with tunics of fine fibers. Stem erect, unbranched.

Leaves five or six, narrowly lanceolate, reaching or slightly exceeding the spike, plane, with a prominent central vein. Spike erect or slightly inflexed, one- to three-flowered; bracts green, $^7/_{16}$–$^{11}/_{16}$ inch (11–18 mm) long, the inner slightly shorter than the outer. Flowers zygomorphic, light orange, probably unscented; perianth tube slender, slightly expanded toward the apex, $^{13}/_{16}$–$^{15}/_{16}$ inch (20–23 mm) long; tepals subequal, lanceolate, $^9/_{16}$–$^{11}/_{16} \times {}^3/_{16}$–$^1/_4$ inch (14–18 × 5–6 mm). Filaments unilateral, exserted about $^9/_{16}$ inch (15 mm) from the mouth of the tube; anthers about $^1/_8$ inch (4 mm) long. Style arching over the stamens, dividing opposite the middle of the anthers, the branches slender, about $^1/_8$ inch (3 mm) long. Capsules subglobose and three-lobed, $^5/_{16}$–$^3/_8$ inch (8–10 mm) in diameter; seeds subglobose, three or four per locule, irregularly globose, apparently $^1/_8$ inch (2.5–3 mm) in diameter Chromosome number unknown. Flowering time mainly late summer to early autumn (February–March).

The only species of *Crocosmia* naturally occurring outside the African continent, *C. ambongensis* is endemic to the Indian Ocean island of Madagascar. Just two populations are known, both from areas of limestone substrate in the western part of the island where the landscape is much eroded by water into steep slopes and narrow gorges, known locally as tsingy. The first was collected in 1903 by the famous explorer of the flora of Madagascar, Henri Perrier de La Bâthie. The second was by a Madagascan nature reserve official in 1954, extending the known range some distance to the south but in the same limestone habitat. Nothing is known about the biology of the species apart from its peculiar habitat. No doubt *C. ambongensis* has a wider range and is not as rare as the current record suggests. The tsingy regions of Madagascar remain poorly collected because of the difficulty of access and the rough terrain that the plant seems to favor. Additional material is needed for further study, and for the present this unusual Madagascan plant remains in *Crocosmia*.

The least well understood species of the genus, *Crocosmia ambongensis* was first assigned to the genus *Geissorhiza* when it was described in 1939. The species does not have the characteristic woody corm tunics of *Geissorhiza,* an exclusively South African genus of some 84 species, all of the winter-rainfall zone of southern Africa. Its placement in *Geissorhiza* was long understood to be incorrect, but because its features seemed most consistent with *Crocosmia* it was finally transferred to that genus (Gold-blatt and Manning 1990a). The species rests there somewhat uncomfortably for it is a small plant with yellow flowers not particularly closely resembling those of the continental African species of the genus. The two Madagascan collections both lack mature capsules and seeds, which are important for distinguishing *Crocosmia* from the related *Tritonia*. The placement of *C. ambongensis* in *Crocosmia* rests largely on leaf anatomical evidence. The species is not known in cultivation.

Crocosmia aurea (Pappe ex J. D. Hooker) Planchon
PLATE 5

> Planchon, Flore des Serres et des Jardins de l'Europe 7: 161 and plate 702 (1851).

> *Tritonia aurea* Pappe ex J. D. Hooker, Curtis's Botanical Magazine 73: plate 4335 (1847). *Babiana aurea* (Pappe ex J. D. Hooker) Klotzsch, Allgemeine Gartenzeitung 1851: 293 (1851). Type: South Africa, without precise locality, as 'George', *K. W. L. Pappe s.n.* (K, holotype; SAM, isotype).

> *aurea,* from the Latin *aureus,* golden, alluding to the deep yellow color of the flowers.

Plants 16–28 inches (40–70 cm) high. Corm almost globose, $^9/_{16}$–$^{13}/_{16}$ inch (15–20 mm) in diameter, the tunics papery, soft-textured, producing long, slender runners as long as 8 inches (20 cm), terminating in new plants. Stem several-branched to unbranched. Leaves sword-shaped, reaching to about the base of the spike, $^3/_8$–1$^3/_{16}$ inches ((1–)2–3 cm) wide, fairly soft-textured, plane with a prominent median vein, the margins hyaline. Spike arching slightly toward the ground, fairly lax, bearing as many as 10 flowers; bracts straw-colored to pale orange, becoming dry as the flowers fade, $^1/_4$–$^9/_{16}$ inch (7–15 mm) long. Flowers actinomorphic, salver-shaped with the tube curved and more or less nodding, orange, pale in the throat, the filaments white, occasionally the tepals marked darker orange to brown above the base, unscented; perianth tube slender, widening slightly toward the mouth, $^3/_8$–$^{13}/_{16}$ inch ((10–)12–20 mm) long; tepals spreading or slightly recurved, $^1/_2$–1$^9/_{16}$ inches × $^1/_4$–$^1/_2$ inch (12–35

(−40) × 6–12 mm). Filaments symmetrically arranged, clasping the style below, weakly diverging above, usually white, $9/16$–$1^3/16$ inches (15–30 mm) long; anthers $1/4$–$3/8$ inch (6–10 mm) long, light orange, pollen orange or yellow. Style directed toward the ground, $7/8$–$1^3/4$ inches (22–45 mm) long, dividing opposite the anthers, the branches weakly diverging, $1/8$–$5/16$ inch (4–8 mm) long. Capsules depressed-globose and three-lobed, showing the outline of the seeds, orange inside; seeds globose, as many as four per locule, about $1/8$ inch (4 mm) in diameter, dark brown to blackish, the coat initially slightly fleshy, becoming wrinkled when dry. Chromosome number $2n = 22$. Flowering time mainly late summer to mid-autumn (February–April in southern Africa but August–October north of the equator).

Crocosmia aurea is typically found on the forest floor and in forest margins, extending from Eastern Cape province, South Africa, to Uganda and Central African Republic. While a successful, and to our collective minds, very attractive garden subject when grown in light shade, in the wild *C. aurea* is most often encountered in the shade of evergreen forest and bush, sometimes in deep shade of the forest floor where the flowers make

Crocosmia aurea

a dramatic splash of bright color. It is an unexpected plant in patches of evergreen African forest where the most common plants on the ground are ferns, species of *Streptocarpus* (Gesneriaceae) and occasionally species of *Aristea* (usually *A. ecklonii* or *A. goetzei*) or *Dietes iridioides*. *Crocosmia aurea* grows on a variety of soils and thrives when provided moisture throughout the growing season.

The flowers are adapted for pollination by large swallowtail butterflies (*Papilio*), perhaps the only organisms that can feed on the nectar held within the slender perianth tube. Although the perianth tube has a relatively wide diameter, access to the tube is restricted to narrow channels between the fairly sturdy, decurrent filaments, which are inserted just inside the tube and held fairly close together, leaving only a tiny space open for access by the thread-like proboscis of a butterfly. Typical of flowers pollinated by butterflies, the nectar has a relatively low concentration of sugars and is rich in glucose and fructose, unlike the nectar offered by Iridaceae pollinated by bees or long-proboscid flies. Butterflies are also probably the only large insects comfortable feeding upside down, for in the nodding *Crocosmia aurea* flowers they must insert the proboscid upward into the tube while hanging upside down from the horizontally extended tepals.

The capsules of *Crocosmia aurea* are also attractive, for the large, shiny, almost black seeds (Plate 5) are displayed on the pale, often orange or even reddish inner surface of the capsules. Together, the clustered seeds resemble a small blackberry and must make a tempting mouthful for a passing bird. Indeed, we think it is likely that the seeds are adapted for dispersal by birds, and this would explain why *C. aurea* seedlings often appear in unexpected places some distance from populations of the species. The reward to birds feeding on the seeds is small, for while the seed coat is moist when fresh, the fleshy layer is thin and, to the human palate, flavorless. Most of the seed consists of a hard, indigestible core.

Crocosmia aurea was first gathered in eastern South Africa by the early plant collector Johann Franz Drège, a pharmacist by trade, in 1832. Drège undertook what today seems a remarkable journey of plant exploration from the eastern part of the Cape Colony, only recently settled by Europeans. The country of the Xhosa people, through which he would have to travel, known today as the Transkei, was beset with unrest. Yet Drège traversed the trackless and dangerous area without incident, arriving some months later at Port Natal, now Durban. His collections, including good specimens of *C. aurea,* were distributed somewhat later and thus were not immediately available to contemporary botanists. While his collections, many of which were the first records of common African plants, attracted much attention when they were distributed to European herbaria, his gathering of *C. aurea* seems to have gone unnoticed.

A decade later, *Crocosmia aurea* was collected in Mozambique, then a Portuguese colony, in the 1840s by Wilhelm Peters. Sent to Berlin after 1848, Peters's collections were examined at the Berlin herbarium, but it was only in the 1860s that the results of Peters's expedition were published. By that time *C. aurea* had been collected again in the Cape Colony, sent to Great Britain, and described. Peters's collection of *C. aurea* was so identified (though with the genus spelled 'Crocosma') in the published account of his expedition, but the manuscript name *Crocanthus mossambicencis,* attributed to J. F. Klotzsch, was also listed in the text. Evidently, Klotzsch (1805–1860) had reached the same conclusion as English and French botanists that the plant represented a newly discovered species and genus. Very likely, Klotzsch's name would have been the one we now use, except for the delayed publication of the account of Peters's travels, *Naturwissenschaftliche Reise nach Mossambique.*

Mystery surrounds the discovery of *Crocosmia aurea* in southern Africa, supposedly at the southern Cape town of George. The species does not grow anywhere nearer than Grahamstown, some 300 miles (500 km) east of George. Pressed plants from George were, nevertheless, received at the Royal Botanic Gardens, Kew, in 1847 together with a painting by the French artist Jean Villet, then living at George. This seems remarkable because in 1847 George was a mere hamlet, geographically and culturally isolated from the nearest center of civilization at Cape Town, 250 miles (400 km) to the west. Plants received at Kew were attributed to Karl Pappe, then Colonial Botanist and resident at Cape Town. Pappe was, however, merely the agent of the shipment of the specimen and painting. We suspect that plants were actually collected close to the eastern frontier of the Cape Colony and brought overland to George from near Uitenhage or Grahamstown. The latter was founded in 1820 by settlers from Britain who prospered and brought British culture, and its associated interests in new plant and animal life, to the Cape's eastern border. No sooner had the specimens reached Kew than Joseph Dalton Hooker, director of the gardens, named the species. He attributed the name *Tritonia aurea* to Pappe, who evidently wrote a provisional name for the species on the specimen label. As a result, not only the name but the gathering of the specimen was for many years attributed to Pappe.

Just 4 years later, plants were being grown in Paris, where Jules Émile Planchon recognized them as the same as those described by Hooker as *Tritonia aurea*. Planchon preferred to regard the species as a member of a new genus, which he called *Crocosmia*, recognizing the faint saffron-like odor of the dried flowers when dampened with water. Saffron is a spice obtained from the orange style branches of *Crocus sativus,* a member of the same subfamily of the Iridaceae as *Crocosmia*. We are always surprised at how rapidly news of scientific discoveries was spread across Europe as early as the middle of the 19th century, even in such apparently obscure subjects as African plants. Thus Planchon knew of Hooker's *T. aurea* and was able to identify the plant he had, from a source not recorded.

Key to *Crocosmia aurea*

1 Stem usually three- or four-branched; spikes exceeding the leaves; perianth tube $^{1}/_{2}$–$1^{1}/_{16}$ inches (13–20(–27) mm) long; tepals $1^{3}/_{16}$–$1^{9}/_{16}$ inches (30–35(–40) mm) long . . . . *C. aurea* subsp. *aurea*
1 Stem usually unbranched, sometimes one- or two-branched; spikes often shorter than the leaves; perianth tube $^{3}/_{8}$–$^{9}/_{16}$ inch (10–12(–15) mm) long; tepals $^{9}/_{16}$–$^{11}/_{16}$ inch (15–18 mm) long
. . . .*C. aurea* subsp. *pauciflora*

We recognize two subspecies of the widespread sub-Saharan *Crocosmia aurea,* the southern African subspecies *aurea* and the tropical African subspecies *pauciflora*. Subspecies *aurea* extends northeast from Uitenhage and Grahamstown in Eastern Cape province, South Africa, through Swaziland, Mozambique, Zimbabwe and Malawi, to northeastern Tanzania. Subspecies *pauciflora* ranges from western Zambia and Angola through Congo, Central African Republic, to Uganda. The two might well be treated as separate species except for intermediate forms where their ranges meet, across central Zambia and Tanzania.

Subspecies *aurea* typically has a multibranched flowering stem, flowers with a tube mostly $^{13}/_{16}$–$1^{1}/_{16}$ inches (20–27 mm) long and tepals mostly $1^{3}/_{16}$–$1^{9}/_{16}$ inches (30–40 mm) long whereas subspecies *pauciflora* has a one- or two-branched or unbranched stem, a perianth tube $^{3}/_{8}$–$^{9}/_{16}$ inch (10–15 mm) long and tepals $^{9}/_{16}$–$^{11}/_{16}$ inch (15–18 mm) long. De Vos (1984), who treated the two as varieties, also stressed the shorter filaments of variety *pauciflora*, about $^{9}/_{16}$ inch (15 mm), and style $^{7}/_{8}$–$1^{1}/_{16}$ inches (22–27 mm) long, whereas subspecies *aurea* has filaments $^{13}/_{16}$–$1^{3}/_{8}$ inches ((20–)25–35) mm long and a style $1^{3}/_{16}$–$1^{3}/_{4}$ inches (30–45 mm) long. We take a very narrow view of subspecies *pauciflora*.

A variety, *maculata*, first recognized by J. G. Baker, raised to species rank by N. E. Brown and treated again as a variety by de Vos, has no taxonomic validity. Defined by the darker orange to brownish markings above the base of the tepals, this is no more than a color variant occasionally seen in the wild, often growing among plants with uniformly colored tepals. It is best recognized as cultivar, 'Maculata' (Plate 5). We have seen occasional plants with marked tepals in populations in the forested valleys inland from Durban in KwaZulu-Natal, and we suspect that similar markings are occasionally found across the range of subspecies *aurea*, though plants with conspicuously marked flowers are rare and we know of no examples of entire populations with boldly marked tepals.

Crocosmia aurea subsp. *aurea*

Crocosmia aurea var. *maculata* Baker, The Gardeners' Chronicle 4: 407, 565 and figure 80 (1888). *Crocosmia maculata* (Baker) N. E. Brown, Transactions of the Royal Society of South Africa 20: 264 (1932). Type: South Africa, without locality, cultivated in Britain, September 1888, from the garden of James O'Brien (1842–1930; Desmond 1994) of Harrow, England, thought to have come from near Algoa Bay, Eastern Cape province (K, holotype).

Crocanthus mossambicencis Klotzsch, mentioned in synonymy by Klatt in W. C. H. Peters, Naturwissenschaftliche Reise nach Mossambique, Botanik 2: 516 (1864), without description and an invalid name.

Flowering stems usually three- or four-branched, occasionally simple. Spikes with 4–10 flowers but the main spike with 8–10 flowers, often shortly exceeding the leaves. Flowers with a perianth tube mostly $^{1}/_{2}$–$^{11}/_{16}$

inches (13–27 mm) long; tepals $1^3/_{16}$–$1^9/_{16}$ inches (30–35(–40) mm) long. Filaments $1^3/_{16}$–$1^3/_8$ inches ((20–)25–35 mm) long; anthers mostly $^1/_4$–$^3/_8$ inch (7–10 mm) long. Style mostly $1^3/_{16}$–$1^3/_4$ inches (30–45 mm) long.

The distribution of *Crocosmia aurea* subsp. *aurea* is virtually the same as that of the entire species, with the exception of western Zambia and adjacent eastern Angola. It thus extends from the Eastern Cape province, South Africa, northward through Swaziland, Zimbabwe, Malawi, Tanzania, to Kenya and Uganda in northeastern Africa and to Cameroon in West Africa.

A most attractive plant, *Crocosmia aurea* subsp. *aurea* is readily distinguished from subspecies *pauciflora* by its branched flowering stem, spikes of as many as 10 large flowers, a perianth tube $^1/_2$–$^{11}/_{16}$ inches ((13–)20–27 mm) long and tepals $1^3/_{16}$–$1^9/_{16}$ inches (30–35(–40) mm) long. The larger perianth is accompanied by larger stamens, the filaments as long as $1^3/_8$ inches (35 mm) and anthers $^5/_{16}$–$^3/_8$ inch (8–10 mm) long. As indicated in the key to the subspecies, flower size readily separates subspecies *aurea* from subspecies *pauciflora*, which has 4–7 flowers per spike, a perianth tube usually $^3/_8$–$^1/_2$ inch (10–15 mm) long and tepals $^9/_{16}$–$^{11}/_{16}$ inch (15–18 mm) long.

Crocosmia aurea subsp. *pauciflora* (Milne-Redhead) Goldblatt

Goldblatt, Flora Zambesiaca 12(4): 50 (1993).

Crocosmia pauciflora Milne-Redhead, Kew Bulletin 1948: 469 (1948). *Crocosmia aurea* var. *pauciflora* (Milne-Redhead) M. P. de Vos, Journal of South African Botany 50: 482 (1984). Type: Zambia, Mwinilunga district, by the Kaoomba River at the edge of evergreen vegetation, 22 December 1937, *Milne-Redhead 3783* (K, lectotype designated by de Vos 1984; K, isotype).

Tritonia cinnabarina Pax, Botanische Jahrbücher für Systematik, Pflanzengeschichte und Pflanzengeographie 15: 152 (1893). *Crocosmia cinnabarina* (Pax) M. P. de Vos, Journal of South African Botany 49: 415 (1983). Type: Angola, Casala Ganginga, January 1881, *Teucsz sub Expedition von Mechow 573* (B, holotype; photo, K).

pauciflora, from the Latin *paucis*, few, and *florus*, flowered, thus few-flowered, referring to the relatively few flowers compared with the branched stems and several-flowered spikes of subspecies *aurea*.

Flowering stems simple or with one or two branches, often not exceeding the leaves. Spike with four to seven flowers. Flowers with a perianth tube $3/8$–$9/16$ inch (10–12(–15) mm) long, tepals $9/16$–$11/16$ inch (15–18 mm) long. Filaments about $9/16$ inch (15 mm) long; anthers about $1/8$ inch (3.5 mm) long. Style $7/8$–$11/16$ inches (22–27 mm) long.

Restricted to western Zambia and adjacent Angola, this small-flowered subspecies of *Crocosmia aurea* is a puzzling plant. The habitat appears to be the same as for *C aurea* subsp. *aurea,* and notes with the type collection tell us that subspecies *pauciflora* grows on the margins of evergreen vegetation in leafy soil and partial shade.

Not nearly as attractive as subspecies *aurea, Crocosmia aurea* subsp. *pauciflora* is readily distinguished by its usually unbranched branched flowering stem (rarely with a short lateral branch), spikes of as many as 4–7 small flowers, a perianth tube $3/8$–$1/2$ inch (10–12 mm) long and tepals $9/16$–$11/16$ inch (15–18 mm) long. The smaller perianth is accompanied by smaller stamens, the filaments usually about $1/2$ inch (15 mm) long and anthers about $1/8$ inch (3.5 mm) long. As indicated in the key to the subspecies, flower size readily separate subspecies *pauciflora* from subspecies *aurea,* which has as many as 10 flowers per spike, a perianth tube $1/2$–$11/16$ inches ((13–)20–27) mm long and tepals $13/16$–$19/16$ inches (30–35(–40) mm) long.

Crocosmia aurea subsp. *pauciflora* was first collected by H. Teucsz, the naturalist attached to Alexandre von Mechow's expedition to interior Angola in 1880–1881. His collection was subsequently named *Tritonia cinnabarina* by the German botanist Ferdinand Pax in 1893. The collection, in fruit, was for a long time misunderstood, and it was not associated with the genus *Crocosmia* until 1983. This was when Miriam de Vos completed her revision of *Tritonia* and after examining the type specimen realized it belonged in *Crocosmia,* to which she transferred the species under the name *C. cinnabarina*. She did not at the time associate the specimen with any known species of the genus.

The plant had meanwhile been discovered in western Zambia in 1937 by the British botanist Edgar Milne-Redhead. Believing that the plant was new to science, he described it in 1948 as *Crocosmia pauciflora*. In her 1984 revision of *Crocosmia* de Vos treated *C. pauciflora* as a variety of *C.*

aurea, being impressed by their similar flowers, those of the former different only in their smaller size.

Crocosmia fucata (Herbert) M. P. de Vos

PLATE 6

M. P. de Vos, Journal of South African Botany 50: 494 (1984).

Tritonia fucata Herbert, Edwards's Botanical Register 24: plate 35 (1838). *Antholyza fucata* (Herbert) Baker, Journal of the Linnean Society 16: 180 (1877b). *Chasmanthe fucata* (Herbert) N. E. Brown, Transactions of the Royal Society of South Africa 20: 274 (1932). *Petamenes fucata* (Herbert) Phillips, Bothalia 4: 44 (1941). Type: South Africa, without precise locality or collector, illustration in Edwards's Botanical Register 24: plate 35 (1838), lectotype designated by de Vos (1984).

fucata, from the Latin *fucatus,* painted or stained, referring to the scarlet to dark red flowers with a yellow throat and darker red, sometimes green, markings on the lower tepals.

Plants 1^1/$_2$–6^1/$_2$ feet (1.3–2 m) high, forming large tufts of as many as 10 plants. Corm globose, 1^3/$_{16}$–2 inches (3–5 cm) in diameter, older corms not resorbed, persisting in a series behind the current corm, reddish. Stem more or less erect, bearing about three cauline leaves flexing outward above the level of the basal leaves, inclined more than 60°, two- to four-branched, the branches diverging 60–80° from the main stem, flushed purplish, about ^{3}/$_{16}$ inch (5 mm) wide below the first branch. Leaves about 10, the lower 7 basal and with long blades, these reaching to about the base of spike, the sheaths forming a false stem below, diverging into a two-ranked fan, the blades lanceolate, 12–16 inches (30–40 cm) long, ^{7}/$_{16}$–1 inch ((11–)20–26 mm) wide, soft-textured, with a prominent central vein, secondary veins all of more or less similar size, the margins not raised. Spike two-ranked, the main axis with as many as 22 flowers, the lateral spikes with as many as 16; bracts membranous below, dry and papery above, the outer about ^{5}/$_{16}$ inch (8 mm) long, acute, the inner slightly longer to slightly shorter than the outer, two-keeled and forked apically as long as ^{1}/$_{32}$ inch (1 mm). Flowers orange-scarlet, sometimes the lower three tepals reddish orange and each with a narrow red median streak, the tube yellow in the underside, unscented; perianth tube trumpet-shaped, curved upward at the base and suberect, $1^9/_{16}$–2 inches (40–50 mm) long, cylindric and slender in the lower $^{13}/_{16}$–1 inch (20–25 mm), gradually expanding and curving outward, the upper about $^{13}/_{16}$ inch (20 mm) forming a wider more or less straight, horizontal tube, about $^1/_4$ × $^1/_8$ inch (6.5 × 4 mm); tepals unequal, lanceolate, the dorsal largest, extended horizontally, about $^5/_8$ × $^5/_{16}$ inch (16 × 8 mm), the upper laterals extending outward at 90° to the tube almost from the base, about $^1/_2$ × $^3/_{16}$ inch (12 × 5 mm), widest at the base, the lower three tepals about $^1/_2$ × $^3/_{16}$ inch (12 × 5 mm), spreading at 90° to the tube. Filaments $1^1/_4$–$1^5/_{16}$ inches (32–34 mm) long, exserted about $^3/_8$ inch (10 mm) from the tube; anthers about $^1/_4$ inch (7 mm) long, dark red, the pollen yellow. Style extending horizontally over the stamens, dividing opposite the upper third of the anthers, the branches about $^1/_{16}$ inch (2 mm) long, minutely notched apically. Capsules subglobose to depressed-globose, three-lobed, the surface rugose to papillate, $^5/_{16}$–$^3/_8$ × $^3/_8$ inch (8–9 × 10 mm); seeds globose, $^1/_8$–$^3/_{16}$ inch (4–5 mm) in diameter, glossy reddish brown, the coat loose and

Crocosmia fucata

78

wrinkled when dry. Chromosome number $2n = 22$. Flowering time late spring to early summer (late October and November).

Restricted to a small part of the area called Namaqualand, the semiarid stretch of country lying along South Africa's western coast and near interior between the Olifants and Orange Rivers, *Crocosmia fucata* has the narrowest range of any of Africa's *Crocosmia* species. It has been recorded on the slopes of Sneeukop (Snow Peak), second highest mountain in the Kamiesberg, and in a valley draining the range to the west on the farm Niekerksfontein. The disjunct distribution of a genus, between the northwestern coast of southern Africa, an area of moderate to low winter rainfall, and summer-rainfall eastern southern Africa and tropical Africa, is surprising but not without parallel. One comparably striking disjunction is in *Clivia,* in the Amaryllidaceae, a genus of four species in forested habitats of eastern southern Africa, and one much more recently discovered species at the southern edge of Namaqualand in the Bokkeveld Mountains (see Manning et al. 2002). Not unexpectedly for the semiarid part of South Africa where it grows, *Crocosmia fucata* favors locally moist habitats and occurs along drainage lines on cool south-facing slopes and stream banks in light shade in gritty, granite-derived loamy soil. The type locality, the farm Niekerksfontein, lies some distance to the west of Sneeukop and at considerably lower elevation. We have not had the opportunity to confirm whether the species still occurs there. The attractive, long-tubed, almost scarlet flowers with yellow markings are adapted for pollination by sunbirds.

Crocosmia fucata has the typical features we associate with *Crocosmia:* a tall, branched flowering stem, branches diverging at a wide angle from the main axis and a strongly arching spike. The long-tubed flowers have a nearly identical structure to those of species like *C. paniculata* and *C. pearsei,* with a narrow cylindric lower portion and a wider, also cylindric upper part. In detail, the flowers can be seen to be somewhat different from those of *C. paniculata,* which has the outer whorl of tepals (the upper laterals and lower median tepals) noticeably smaller than the inner three and a slight but unmistakable pouch where the lower part of the perianth tube expands into the wider upper portion. The flowers of *C. fucata* are predominantly scarlet with the lower three tepals each with a narrow red median streak and the tube yellow in the underside and in the throat. The Namaqualand species, however, differs sharply from its apparent relatives in eastern southern Africa in having plane, fairly narrow leaves whereas *C. paniculata* and *C. pearsei* have broader, strongly pleated leaf blades that taper markedly at the base and tip. The distinctive long-tubed flowers of those two species are adapted for pollination by long-billed sunbirds, and it is possible that this flower form evolved independently in *C. fucata* and the two other long-tubed species of the genus.

As related in the chapter, Early Exploration and the Discovery, *Crocosmia fucata* first came to botanical attention in 1838 when a painting of a plant in full bloom was published in Sydenham Edwards's *Botanical Register* together with a formal botanical description by William Herbert, the British bulbous plant specialist. Herbert, who named the species *Tritonia fucata,* relates that he had grown the plant for some 25 years before it flowered in his garden the previous year for the first time. That means that it reached Britain no later than 1812. There appears to be no record of the original collection grown by Herbert, and *C. fucata* remained for a long time unknown in the wild. An interesting illustration of the species in Mrs Jane Loudon's 1841 book, *The Ladies' Flower-Garden of Ornamental Bulbous Plants,* is copied from the painting published by Herbert in 1838.

The German plant explorer Carl Ludwig Zeyher must be credited with the first record of *Crocosmia fucata* in the wild. Plants he collected, evidently in 1829 (Gunn and Codd 1981) at 'Nieuw Kerksfonteyn' at 1200–1500 feet (370–460 m), now the farm Niekerksfontein, were not at first associated with *C. fucata.* It was only in 1910 that the range of *C. fucata* was finally confirmed when the expedition led by first director of the National Botanical Garden at Kirstenbosch, Henry H. W. Pearson, found plants on the upper slopes of Sneeukop in the Kamiesberg of Namaqualand. J. G. Baker transferred *Tritonia fucata* to *Antholyza* in 1877, and in 1932 N. E. Brown removed it to *Chasmanthe.* It only found its position in *Crocosmia* in 1984 after M. P. de Vos examined living specimens re-collected plants on the farm Modderfontein, on the eastern slopes of Sneeukop. The plant illustrated here is from the same place, where we found *C. fucata* growing in quantity in 1995.

It is puzzling that *Crocosmia fucata* is not in cultivation. The flowers

The Plates

by Auriol Batten

Plate 1.
Chasmanthe aethiopica

Plate 2.
*Chasmanthe
bicolor*

Plate 3. *Chasmanthe floribunda,* inflorescence of the yellow variant sometimes called variety *duckittii* toward the left

Plate 4. *Crocosmia ambongensis,* from photographs of herbarium specimens at the Natural History Museum, Paris

1 cm

Plate 5. *Crocosmia aurea,* flowers of the variant with darker markings above the base of the tepals, 'Maculata', toward the middle

Plate 6.
Crocosmia fucata

Plate 7.
*Crocosmia
masoniorum*

Plate 8.
*Crocosmia
mathew-
siana*

Plate 9.
*Crocosmia
paniculata*

Plate 10.
*Crocosmia
pearsei*

Plate 11.
*Crocosmia
pottsii*

Plate 12.
Crocosmia
×crocosmiiflora

Plate 13.
Crocosmia cultivars. Clockwise from upper left: 'Star of the East', 'Lucifer', 'Vulcan', 'Sir Matthew Wilson', 'Lady Hamilton', 'Croesus'

Plate 14.
Crocosmia cultivars. Clockwise from left: 'Jackanapes', 'His Majesty', 'Queen Alexandra', 'Solfatare', 'Comet', 'Nimbus'

Plate 15. *Crocosmia* cultivars. Clockwise from upper left: 'Mrs Geoffrey Howard', 'Mephistopheles' (center), 'Culzean Pink' (upper right), 'Gerbe d'Or', 'Prometheus', 'Emily McKenzie'

are striking, and the fact that it could be flowered in open ground in York-shire in the British Isles suggest that it would make a valuable addition to the horticultural palette. It is winter growing and flowers in late spring or early summer, which makes it appear an ideal subject for gardens in areas with a Mediterranean climate. In the wild, plants begin to put out new leaves in late March or April, often before the first rains of the season have fallen but after night temperatures have dropped and the days have become shorter.

Crocosmia masoniorum (H. M. L. Bolus) N. E. Brown
PLATE 7

> N. E. Brown, Transactions of the Royal Society of South Africa 20: 264 (1932). De Vos, Journal of South African Botany 50: 484 (1984).

> *Tritonia masoniorum* H. M. L. Bolus, Annals of the Bolus Herbarium 4: 43 (1926), as '*masonorum*'. Type: South Africa, Eastern Cape, Tembuland, Engcobo Mountain, 4500 feet (1400 m), 20 January 1896, *H. Bolus 10303* (BOL, lectotype designated by de Vos 1984; K, isolectotype).

> *masoniorum,* Latin adjective from the English surname Mason (*Masonius* in Latin), the plural genitive being *masoniorum,* thus named in honor of both Marianne Mason and her brother Edward, who made an important early collection of the species.

Plants mostly 20–30 inches (50–80 cm) high. Corm globose, persisting, thus superposed on one another, forming chains, about 1 inch (25 mm) in diameter, lateral cormlets usually numerous, $3/8$–$9/16$ inch (10–15 mm) in diameter, the tunics brown, fibrous, soon disintegrating. Stem suberect or inclined to drooping, simple or with one or two short branches. Leaves more or less lanceolate, tapering to a narrow, petiole-like base, pleated, usually shorter than the stem, 1–2 inches (25–50 mm) wide, decreasing in size above. Spike curved outward at the base and more or less horizontal, with many flowers all borne on the upper side and facing the spike apex; bracts soft-textured, green, becoming dry from the tip, the outer $3/16$–$3/8$ inch (5–10 mm) long, the inner bifid and shorter than the outer. Flowers zygomorphic, brilliant orange-scarlet, rarely yellow, the tepals spreading when fully open, drooping slightly at the tips, unscented; perianth tube obliquely funnel-shaped, curving upward, $^{11}/_{16}$–1 inch (18–25 mm) long, the lower half narrow and cylindric, widening above, as wide as $^3/_8$ inch (10 mm) at the mouth; tepals unequal, the dorsal somewhat larger, $^{13}/_{16}$– $1^3/_{16}$ inches (20–30 mm) long, suberect, the others $^{11}/_{16}$–1 inch (18–26 mm) long, curving outward, the outer three $^3/_{16}$–$^1/_4$ inch (5–7 mm) wide, the inner $^1/_4$–$^1/_2$ inch (7–12 mm) wide. Filaments $1^3/_{16}$–$1^3/_8$ inches (30–35 mm) long; anthers $^1/_4$–$^3/_8$ inch (7–9 mm) long, reaching to the top of the dorsal tepal. Style well exserted from the tube, $1^5/_8$–$1^3/_4$ inches (42–45 mm) long, dividing opposite the anther tips, the branches about $^1/_8$ inch (3.5 mm) long, bilobed to forked at the apices. Capsules depressed-globose and three-lobed, about $^3/_8 \times ^3/_8$ inch (10 × 9 mm), glossy when still green; seeds about $^1/_8$ inch (4 mm) in diameter, ovoid-angular, lightly wrinkled, dark brown, two or three per locule. Chromosome number $2n = 22$. Flowering time early summer to midsummer (December–January).

Crocosmia masoniorum

The rare *Crocosmia masoniorum* is known from only a handful of localities, all in the mountains that lie to the north of the towns of Engcobo and Umtata in Transkei, the region of South Africa's Eastern Cape province that lies between the Kei River to the south and the KwaZulu-Natal border to the north. Although records of the species are few, we suspect that it occurs in suitable moist, rocky habitats all along the edge of the middle Drakensberg escarpment from Cala Pass in the west to Nquadu (north of Umtata) in the east. Access to this extensive tract of largely roadless country of deep valleys and steep mountain slopes is difficult, making the area still today relatively poorly collected though its rich flora has much to offer the adventurous plant hunter. Not surprisingly, most records are from places where roads cut inland, at Cala Pass and Satan's (also called Satana's) Neck. Plants grow in rocky sandstone outcrops, often on cliffs, or on stream banks. The appearance of the flowers, with a relatively short perianth tube and weak, drooping flowering stem, suggests that they are adapted for pollination by large butterflies. In fact the general aspect of the entire flowering spike with is crowded flowers is reminiscent of the Cape species *Gladiolus nerineoides,* a plant that also grows on cliffs and steep rocks and is pollinated by a large butterfly, in this case the Cape Christmas butterfly, also called the pride of Table Mountain, *Aeropetes tulbaghia.*

De Vos (1984) considered *Crocosmia masoniorum* to be most closely related to pleated-leaved *C. mathewsiana* of Mpumalanga province in South Africa, to the north of KwaZulu-Natal, but the much taller *C. mathewsiana* has smaller, shorter-tubed flowers, short stamens with the filaments $1/2$–$9/16$ inch (13–15 mm) long and a much-branched flowering stem with the spikes nodding in bud. Although the two species are clearly allied, a closer relationship with the high-Drakensberg *C. pearsei* seems more likely. That species also has a few-branched, relatively short flowering stem and pleated leaves tapering below into a narrow, petiole-like base and favors steep slopes and cliffs, but the flowers have a longer tube, $1 9/16$–$1 3/4$ inches (40–45 mm) versus $11/16$–1 inch (18–25 mm) in *C. masoniorum.*

Although not widely known in cultivation, *Crocosmia masoniorum* is easy to grow and makes a remarkable display in December and January in gardens in the Southern Hemisphere or in July and August in the Northern Hemisphere. The wild species is a desirable garden subject, and the yellow variant that arose spontaneously in cultivation at Rowallane Garden in Ireland, called 'Rowallane Yellow', is even finer, with a strong, clear yellow color. The yellow-flowered form has evidently arisen repeatedly for it is also found in South African gardens where is known as the golden swan crocosmia. We suspect that yellow-flowered plants may occur sporadically in the wild. The yellow-flowered variant of *C. masoniorum* is paralleled in yellow-flowered sports of *C. paniculata* recorded from Mpumalanga province, occasional yellow-flowered individuals of *Chasmanthe aethiopica,* and by the now widely cultivated yellow-flowered variant of *C. floribunda.* The latter plant has been named variety *duckittii* but we suggest it should be regarded as a cultivar.

Crocosmia masoniorum was named in honor of the artist Marianne H. Mason and her brother Edward Mason, of Saint Bede's College in Umtata. They collected plants in 1911 in the northeastern part of South Africa's Eastern Cape province, an area known as the Transkei, between Engcobo and the trading post of Zidungeni 'on a steep track overhanging a little waterfall in a tiny kloof, with tree ferns and other ferns, a few yards, perhaps 50, below the road.' Correspondence preserved at the Kew herbarium from Miss Mason to 'Mr [? N. E.] Brown', and an article in the *Journal of the Royal Horticultural Society* (Mason 1913), tells us that this is exactly the same place that the species was first collected in January 1896 by Harry Bolus.

Plants still grow there today in this very attractive setting, but sadly, the alien Australian wattle *Acacia mearnsii* is spreading into the valley. If not checked, this aggressive tree species will certainly alter the habitat and probably eliminate *Crocosmia masoniorum* at its type locality. Plants were introduced into cultivation about 1912, at the Cambridge University Botanic Garden, from corms brought from South Africa by Marianne Mason. This original introduction of the species, then known only as a species of *Tritonia,* evidently did not survive. The source of later introductions to Europe is apparently not recorded.

Perhaps the loveliest of all *Crocosmia* species, *C. masoniorum,* the swan crocosmia, is less well known in cultivation than it deserves. Perhaps this is due in large measure to difficulty in obtaining stock, for it is relatively easy to grow. It appears to grow as successfully in such different climates as that of the South African high veld, with its dry, frosty winters, and Ireland, with a damp climate and cool summers. *Crocosmia masoniorum* is occasionally seen for sale as a cut flower in the western Europe and the United States. It received an Award of Merit from the Royal Horticultural Society in 1957.

Although often spelled *Crocosmia 'masonorum'*, as in the protologue by the South African botanist Louisa Bolus and in N. E. Brown's (1932) and M. P. de Vos's (1984, 1999b) accounts of the species, the correct Latin form of 'Mason' is *Masonius*, hence *C. masoniorum*. Like most other species of *Crocosmia, C. masoniorum* was first referred to *Tritonia* (H. M. L. Bolus 1926). It was transferred to *Crocosmia* by N. E. Brown in 1932.

Crocosmia mathewsiana (H. M. L. Bolus) Goldblatt ex de Vos
PLATE 8

M. P. de Vos, Journal of South African Botany 50: 488 (1984).

Tritonia mathewsiana H. M. L. Bolus, Annals of the Bolus Herbarium 3: 76 (1921). Type: Mpumalanga (as 'Transvaal'), Pilgrim's Rest district, near Graskop, flowered at Kirstenbosch in 1916, *Wood s.n.* (holotype, National Botanical Garden 542/16, BOL).

mathewsiana, Latin epithet honoring the curator of the National Botanical Garden, Kirstenbosch—Joseph William Mathews—who worked closely with the botanist H. M. L. (Louisa) Bolus in growing many newly discovered species for illustration and additional study.

Plants 5–8 feet (1.5–2(–2.5) m) high. Corm globose, $1^3/_{16}$–2 inches (3–5 cm) in diameter, those of the past two or three seasons usually persisting under the current one, the tunics of unbroken papery layers, becoming fibrous with age. Stem erect or inclined, four- to seven-branched, the lower branches also sometimes once- or twice-branched. Leaves several, lanceolate, pleated, firm-textured, reaching to about the base of the spike, as wide as $1^3/_8$ inch (35 mm). Spikes drooping in bud, becoming horizontal as the flowers open, two-ranked, the main spike 18- to 30-flowered, lateral spikes with fewer flowers; bracts firm-textured, purple below, the apices dry and brown, apiculate or obscurely trifurcate, $^3/_{16}$–$^1/_4$ inch (5–7 mm) long, the inner bracts usually slightly shorter than the outer,

$^3/_{16}$–$^1/_4$ inch (5–6 mm) long, forked at the tip. Flowers zygomorphic, uniformly bright orange, facing to the side or half-nodding, unscented; perianth tube narrowly trumpet-shaped, $^5/_8$–$^{13}/_{16}$ inch (16–20 mm) long, slightly arched; tepals unequal, nearly symmetrically disposed (the dorsal slightly larger), inner tepals largest, oval-elliptic, $^9/_{16}$–$^5/_8$ × about $^3/_8$ inch (15–16 × about 9 mm), outer tepals narrowly deltoid, about $^9/_{16}$ × $^3/_{16}$ inch (14 × 5 mm). Filaments $^1/_2$–$^9/_{16}$ inch (13–15 mm) long, exserted about $^1/_8$ inch (3 mm) from the tube; anthers parallel and contiguous, $^1/_4$ inch (6–7 mm) long, pale yellow, the pollen yellow. Style arching behind the filaments, $^{13}/_{16}$–$^{15}/_{16}$ inch (20–23 mm) long, dividing just below the anther bases, the branches $^3/_{16}$–$^3/_8$ inch (5–10 mm) long, forked apically, laxly spreading. Capsules depressed-globose and three-lobed, $^1/_4$ inch (6–7 mm) long, about $^5/_{16}$ inch (8 mm) in diameter; seeds angular-prismatic, dark red-brown, about $^1/_8$ inch (3 mm) long, mostly one or two per locule, rarely three. Chromosome number unknown. Flowering time summer to early autumn (late January to March).

Crocosmia mathewsiana

Locally common, *Crocosmia mathewsiana* is a narrow endemic of Mpumalanga province, South Africa. It is restricted to the lower escarpment of the northern Drakensberg between Graskop and Mariepskop, near Blyde River Canyon. Plants grow in light shade along the edges of elfin forest, sometimes within the forest, or at the edges of taller, evergreen forest. Plants form scattered populations. The species occurs in well-drained sandy ground among boulders in the geological formation known as Black Reef Quartzite.

The flowers are adapted for pollination by large anthophorine bees, particularly species of *Amegilla* (family Apidae). These active, long-tongued bees have a body that fits snugly into the upper part of the perianth tube so that their tongues, as long as $1/2$ inch (12 mm), can be inserted into the narrow lower part of the tube, which contains nectar. Pollen is brushed against the upper part of the thorax as the bee forages for nectar and is in turn transferred to stigmatic surfaces of the next flower visited.

Crocosmia mathewsiana is readily identified by the combination of strongly pleated leaves, multibranched flowering stem, drooping, flexuose spikes and relatively short-tubed, brilliant orange-scarlet flowers. We assume that leaf pleating evolved just once in the genus, so the relationships of *C. mathewsiana* must lie with the three other pleated-leaved species of the genus. Its seems most likely that it is most closely related to *C. paniculata,* which also has a multibranched flowering stem and strongly flexuose spikes. In fact, as de Vos (1984) commented, the two species can barley be distinguished from one another in the vegetative state, though the stem and branches of *C. mathewsiana* are more slender and strongly nodding in bud, and the spikes less prominently flexuose. The flowers of the two species, however, differ considerably, for those of *C. paniculata* have an elongate perianth tube, $1^5/_{16}$–$1^9/_{16}$ inches (33–40 mm) long, versus about $^{13}/_{16}$ inch (20 mm) in *C. mathewsiana,* and are adapted for pollination by sunbirds in contrast to bee pollination in *C. mathewsiana.*

Crocosmia mathewsiana was evidently first collected in February 1874 'in the vicinity of Lydenburg' in what was then the South African Republic, now Mpumalanga province of South Africa, by one W. Roe, about whose botanical activities nothing seems to have been recorded. This early collection, housed at the herbarium of the Royal Botanic Gardens,

Kew, was associated with *C. pottsii* and elicited no immediate attention despite what now seems obvious, that it was a different species quite unlike *C. pottsii* in its broad, pleated leaves and compact, dropping spikes. In fact, it was only identified as *C. mathewsiana* by de Vos in 1984. A later collection from Graskop, near Pilgrim's Rest, not far from Lydenburg, was sent to Kirstenbosch, where it flowered and was duly described by Louisa Bolus in 1923. We know nothing about the original collector, a Mr Wood (not the famous Natal botanist John Medley Wood), but Bolus named the species in honor of the curator of the garden, Joseph William Mathews. Assigned to *Tritonia* when described, the species was only transferred to *Crocosmia* by de Vos in her revision of the genus in 1984. A chromosome count made by Goldblatt (1971), treated the species as *C. mathewsiana*, but the combination had, in fact, not been validly published at the time. De Vos (1984) noted that the voucher documenting the identity of the plant is actually *C. paniculata,* hence the chromosome number of *C. mathewsiana* remains to be determined.

Crocosmia paniculata (Klatt) Goldblatt

PLATE 9

> Goldblatt, Journal of South African Botany 37: 444 (1971). De Vos, Journal of South African Botany 50: 490 (1984).

> *Antholyza paniculata* Klatt, Linnaea 35: 379 (1867). *Curtonus paniculatus* (Klatt) N. E. Brown, Transactions of the Royal Society of South Africa 20: 270 (1932). Type: South Africa, without precise locality, as 'Natal and Zululand', about 1864, *W. T. Gerrard 1530* (K, holotype; BOL, NH (as *W. T. Gerrard and M. J. McKen*), isotypes).

> *paniculata,* from the botanical term panicle, a loosely and highly branched inflorescence (or cluster of flowers), derived from the Latin *panum,* an ear of millet, which consists of a multibranched, almost brush-like cluster of flowers.

Plants mostly 4–5 feet (1.2–1.5 m) high. Corm globose, persisting, often superposed on one another, forming short chains, $1^3/_{16}$–$1^3/_8$ inches (30–35 mm) in diameter, the tunics brown, papery or becoming fibrous with age. Leaves more or less lanceolate, pleated, reaching to the base of the spikes, decreasing in size above, mostly 1³⁄₈–2 inches (35–50 mm) wide. Stem erect or somewhat inclined, with three to six diverging branches, these sometimes also branched, usually dull purplish red. Spike arching outward and more or less horizontal, with 18–25 flowers borne on the upper side in two rows facing away from one another; bracts firm-textured, purplish green, becoming dry from the tips, the outer ¼–³⁄₈ inch (6–10 mm) long, sometimes obscurely trifid, the inner two-veined, forked at the tip, slightly shorter than the outer. Flowers zygomorphic, deep orange to orange-brown, sometimes shaded with reddish brown outside, yellowish inside the tube, the lower three tepals each with marked in the lower third with a dark red median streak decurrent on the tube, more or less trumpet-shaped, the dorsal tepal extending horizontally, remaining tepals spreading distally, unscented; perianth tube narrow below, abruptly widened into a cylindric upper part, curved in the middle, mostly 1⁵⁄₁₆–1⁹⁄₁₆ inches (33–40 mm) long, sometimes as long as 2 inches (50

Crocosmia paniculata

mm), the upper part $9/16$–1 inch (15–25 mm) long; tepals unequal, the dorsal largest, inclined to nearly horizontal, $9/16$–$11/16$ × $5/16$–$3/8$ inch (14–18 × 8–10 mm), united with the upper laterals for about $3/16$ inch (5 mm), upper laterals narrowly deltoid, about $3/8$ × $3/16$ inch (10 × 5 mm), lower lateral tepals narrowly ovoid to oblong, $7/16$–$9/16$ × $3/16$–$1/4$ inch (11–14 × 5–6 mm), lower median narrowly deltoid, directed downward, about $3/8$ × $1/8$ inch (10 × 4 mm). Filaments 1–$1^9/16$ inches (25–40 mm) long, exserted $1/2$–$11/16$ inch (12–18 mm) from the tube; anthers parallel and contiguous, about $1/4$ inch (6 mm) long, dull red or yellow with red on the lines of dehiscence, the pollen yellow. Style well exserted from the tube, mostly 2–$2^9/16$ inches (50–65 mm) long, dividing between the base and middle of the anthers, the branches $3/16$–$1/4$ inch (5–7 mm) long, forked apically for less than $1/32$ inch (1 mm), diverging and arching forward, shortly exceeding the anther apices, bilobed to forked at the apices. Capsules depressed-globose and three-lobed, about $1/4$ × $5/16$ inch (5.5 × 8 mm), flushed dull red to purple, slightly rough to the touch; seeds globose or the sides slightly flattened by pressure, lightly wrinkled, ovoid-angular, one to three per locule, about $1/8$ inch (3 mm) long, dark red-brown. Chromosome number $2n = 22$. Flowering time mainly early summer to late summer (December to mid-February).

Native to eastern southern Africa, *Crocosmia paniculata* extends from the northern Mpumalanga highlands near Blyde River Canyon and Graskop in South Africa in the north through Swaziland, interior Mpumalanga and northeastern Free State into Lesotho and locally in the foothills of the KwaZulu-Natal Drakensberg (for example, the 1886 collection *J. M. Wood 3499,* from the banks of Klip River). The southernmost station so far recorded is the upper Umzimkulu Valley, made surprisingly as early as 1895, by the naturalist Alice Pegler. It probably also occurs in extreme northern KwaZulu-Natal, adjacent to Mpumalanga, but that awaits confirmation. *Crocosmia paniculata* is also recorded from the eastern highlands of Zimbabwe where the first record from that country dates from 1955. This relatively late date suggests that it is not native there. There is an unexpected gap in its range between eastern Zimbabwe and the northern Mpumalanga highlands, for *C. paniculata* does not occur in the eastern highlands of Limpopo province, where there is ample suitable habitat for the species. Plants favor moist situations and may be found along streams, drainage lines and in marshes.

The bright to dull orange or reddish flowers of *Crocosmia paniculata* have a perianth tube $1^5/_{16}$–$1^9/_{16}$ inches (33–40 mm) long that contains nectar in the narrow lower portion, accessible only to long-billed sunbirds. While there are few published records of floral visitors to the species, it is not unusual to see sunbirds perched on the strong branches, feeding on the nectar. We have seen the double-collared sunbird, *Nectarinia afra,* visiting *C. paniculata* near Graskop in Mpumalanga province. We suspect other species of sunbirds also visit and pollinate the flowers.

Plants we consider typical *Crocosmia paniculata* have bright orange to brownish orange flowers, also sometimes described as scarlet, with a perianth tube $1^5/_{16}$–$1^9/_{16}$ inches (33–40 mm) long and a more or less horizontal dorsal tepal $^9/_{16}$–$^{11}/_{16}$ inch (15–18 mm) long, well exceeding the lateral and lowermost tepals, which are only $^1/_4$–$^1/_2$ inch (6–12 mm) long. The tallest species of the genus, *C. paniculata* may reach 5 feet (1.5 m) in height in suitable habitats and has a very handsome appearance. Particularly striking, the repeatedly branched flowering stem forms a compound arrangement of horizontal, strongly flexuose spikes on which the flowers are arranged in two rows, the flowers facing away from one another. The closely related *C. pearsei,* until 1981 included in *C. paniculata,* occurs in the high Drakensberg and is a smaller, few-branched plant with even larger, longer-tubed flowers than those of *C. paniculata.* They have a perianth tube $1^9/_{16}$–$1^3/_4$ inches (40–45 mm) long and a dorsal tepal about $^{13}/_{16}$ inch (20 mm) long. *Crocosmia pearsei* grows in a habitat quite different from that of *C. paniculata,* favoring basalt rock outcrops, steep banks and cliffs. The Namaqualand species *C. fucata* also has flowers that closely resemble those of *C. paniculata,* and the species may be closely related. The two differ in two important features: leaves and the flowering stem. Leaves of *C. fucata* have a flat blade of somewhat softer texture than in *C. paniculata,* and the flowering stem has at best only one or two branches, quite unlike the highly branched flowering stem in *C. paniculata.*

Mystery surrounds the first collection and type of *Crocosmia paniculata.* It was discovered by the early KwaZulu-Natal collector William Tyrer Gerrard, who sent specimens to the Royal Botanic Gardens, Kew, in 1865. Data on a duplicate of the type collection at the KwaZulu-Natal Herbarium in Durban, South Africa, attribute the collection to Gerrard and Mark Johnston McKen, the first curator of the Natal Botanic Garden there. This is not unexpected, for Gerrard and McKen made several collecting trips together in the 1860s. The type specimen at the herbarium of the Royal Botanic Gardens, Kew, records the date received, July 1865, but not the date collected, and the locality, either 'Natal and Zululand' on the Kew specimen, or simply 'Natal', is unsatisfactory, because plants matching the type collection are rare in KwaZulu-Natal, though they occur immediately adjacent to it northern and northeastern borders in Free State and Mpumalanga provinces, near Harrismith and at Oshoek near Wakkerstroom, for example.

Confusion has surrounded the identity of *Crocosmia paniculata* because the common *Crocosmia* seen today in the KwaZulu-Natal Midlands, for example, near Hilton, Howick and Nottingham Road, has dark reddish flowers that closely resemble those of *C. paniculata* except in their color and the shorter perianth tube, only 1–1 $^1/_8$ inches (26–28 mm) long. These red-flowered plants are never seen in truly wild situations; they are either cultivated in gardens or found near abandoned dwellings, road verges and ditches. There are evidently no collections of this plant made before 1900, by which time KwaZulu-Natal had been widely botanized, and such a conspicuous plant would surely have been recorded. In fact, the first record of the red-flowered, short-tubed plant appears to be one collected at Pietermaritzburg, capital of the province, by J. Medley Wood in 1909.

We suggest that the red-flowered plant may be an old cultivar no longer available in the trade, one well adapted to the warm, well-watered KwaZulu-Natal climate. The plants are fertile and as far as known they breed true from seed, although this should be critically confirmed. To call it *Crocosmia paniculata* is mistaken, but so widespread is the belief that this plant represents the type form that identifications in the literature of any plant as *C. paniculata* must be suspect.

Crocosmia paniculata was described in 1867, soon after its discovery, by the German specialist of the Iridaceae, F. W. Klatt. Mpumalanga province was explored botanically much later than KwaZulu-Natal, but is noteworthy that the British horticulturist and traveler Christopher Mudd, sent to South Africa by the famous nursery James H. Veitch and Sons, of Chelsea, England, collected plants on the eastern escarpment at MacMac

Falls near Sabie in the summer of 1877–1878. Nearby, Mudd also made the first collection of the fine, large, pale yellow-flowered *Moraea* (also Iridaceae) that was named in his honor, *M. muddii,* by N. E. Brown in 1929. But his collection of *C. paniculata* elicited no such scientific attention. Whether Mudd collected seeds or corms as well as preserved material is not known, but we assume that he did as he was employed by a nursery that presumably wished to obtain new plants for sale.

Crocosmia paniculata is first recorded in cultivation in the British Isles, where in 1884 a Mr Miles, who gardened at Royston, England, sent flowers to the horticultural magazine *The Garden.* He had collected the species in the Transvaal (then the South African Republic) and had grown it outdoors for some years, where it increased rapidly. The 'long panicle of brown red and yellow flowers' would appear to confirm that Miles had the race of *C. paniculata* that grows in what is now northern Mpumalanga province. The note, written by editorial staff of *The Garden,* was published 25 October 1884 (Robinson 1884).

Crocosmia pearsei Obermeyer

PLATE 10

> Obermeyer, Bothalia 13: 450 (1981). De Vos, Journal of South African Botany 50: 493 (1984). Type: South Africa, KwaZulu-Natal, Drakensberg, summit of Mnweni Pass, 1 April 1977, *R. O. Pearse 34* (PRE, holotype).

> *pearsei,* named for Reginald Oliver Pearse, who was responsible for drawing attention to this neglected species, which until the publication of his book, *Mountain Splendour,* was confused with *Crocosmia paniculata.*

Plants mostly 31–40 inches (80–100 cm) high. Corm globose, persisting, thus superposed on one another, forming chains, about 1 inch (25 mm) in diameter, the tunics brown, fibrous. Stem suberect or inclined to horizontal, simple or with one or two short branches. Leaves more or less lanceolate, tapering to a narrow, petiole-like base, pleated, shorter than the stem, decreasing in size above, $1^3/_8$–$2^3/_4$ inches (35–70 mm) wide. Spike arching outward and more or less horizontal, with many flowers all borne on the upper side and forming two rows facing away from one another;

bracts soft-textured, purplish green, becoming dry from the tips, the outer $5/16$–$3/8$ inch (8–10 mm) long, the inner bifid and shorter. Flowers zygomorphic, deep orange, more or less trumpet-shaped, the dorsal tepal extending horizontally, remaining tepals arching outward and spreading in the upper half, unscented; perianth tube narrow below, wider and cylindric above, $1^{9/16}$–$1^{3/4}$ inches (40–45 mm) long; tepals unequal, the dorsal largest, $13/16$–$1^{3/16}$ inches × $3/8$–$1/2$ inch (20–30 × 10–12 mm), remaining tepals $9/16$–$13/16$ × $1/4$–$5/16$ inch (14–20 × 7–8 mm). Filaments $1^{3/4}$–2 inches (45–50 mm) long; anthers $1/4$ inch (6–7 mm) long. Style well exserted from the tube, 2–$2^{3/8}$ inches (50–60 mm) long, dividing opposite the upper half of the anthers, ultimately the branches exceeding the anther tips, about $1/8$ inch (3.5 mm) long, bilobed to shallowly forked at the tips. Capsules depressed-globose and three-lobed, $1/4$ × $5/16$ inch (6 × 8 mm), green, often flushed dull red, drying brown; seeds more or less globose, dull reddish brown. Chromosome number unknown. Flowering time mainly mid- to late summer (December to early February).

Crocosmia pearsei

Originally known from a few inaccessible sites in the high Drakensberg of southern Africa, near Cathedral Peak in KwaZulu-Natal province, South Africa, records now show that *Crocosmia pearsei* extends to the north as far as Basuto Gate, north of The Sentinel in Free State province, and almost certainly also grows in adjacent eastern Lesotho. We have seen plants flowering fairly close to the road near Witzieshoek Mountain Resort and being visited there by sunbirds, their pollinators. Plants grow in basalt outcrops and cliffs, mostly on cooler slopes at elevations above 7000 feet (2000 m). Most populations are difficult to reach because of their steep habitat, but occasionally, as at Basuto Gate, one can walk right up to them. While this makes it easier to examine and collect plants, they suffer from browsing by antelope and accessible plants almost always have their apparently tasty fruiting spikes eaten off. *Crocosmia pearsei* does not appear to respond well to cultivation, most likely because of its specialized high-mountain habitat, preference for icy cold winters, cool summers and well-drained soil.

Until 1981, the few collections of *Crocosmia pearsei* was confused with the closely related *C. paniculata,* and it was only after the amateur naturalist R. O. Pearse pointed out their difference that it became clear that this is a separate species. Plants can be recognized by their smaller stature, seldom reaching 1 m in height, and the stems are often unbranched or at best have one or two short lateral branches, unlike the strongly branched stems of *C. paniculata.* The flowers of *C. pearsei* are, however, even larger than those of its more common relative. Some $2^3/8$–$2^3/4$ inches (60–70 mm) long, they have a perianth tube $1^9/16$–$1^3/4$ inches (40–45 mm) long and a dorsal tepal about $^{13}/16$ inch (20 mm) long. In contrast, what we consider typical *C. paniculata* has orange to brownish orange flowers with a perianth tube $1^5/16$–$1^9/16$ inches (33–40 mm) long and a more or less horizontal dorsal tepal $^1/2$–$^{11}/16$ inch (14–18 mm) long. The spikes are strongly flexuose, and the flowers are borne in two opposed rows, facing away from one another. The repeatedly branched flowering stem of *C. paniculata* forms a compound panicle-like inflorescence quite different from the simple spike of *C. pearsei.* De Vos (1984) commented that *C. pearsei* might be considered a local variant of *C. paniculata,* but we cannot agree. The two species not only differ in the vegetative and floral features just listed but grow in quite different habitats, *C. paniculata* being a plant of wet habitats, where it grows in deep, rich soils at moderate elevations, unlike the high-altitude, well-drained basalt slopes and cliffs where *C. pearsei* is found.

One unusual feature of *Crocosmia pearsei* is the chains of small corms, which persist below the current corm. This feature is matched in the genus by another pleated-leaved species, the southern Drakensberg *C. masoniorum.* This similarity suggests a possible relationship between the two, though the flowers of *C. masoniorum* (Plate 7) are rather different; they are borne on an unbranched or few-branched stem and in a single row on the horizontal spike.

The last species of *Crocosmia* to be described, *C. pearsei* had been known for many years but was confused with the similar *C. paniculata,* the flowers of which seem at first to be identical. The late Reginald Oliver Pearse, teacher and conservationist, intensely interested in the Drakensberg flora, was the first to conclude that the high-elevation Drakensberg plants referred to *C. paniculata* are distinct from that species. Photographs in his book, *Mountain Splendour,* published in 1978, demonstrate convincingly the difference between the two species. The Drakensberg plant was appropriately named in his honor by the South African botanist Amelia Obermeyer (Mrs A. A. Mauve) in 1981.

Crocosmia pearsei was first introduced into cultivation in Great Britain in 1991 by James Compton, John D'Arcy and Martyn Rix. Normally a solitary plant in the wild, it increases vegetatively quite slowly in cultivation. At the Royal Botanic Garden, Edinburgh, it has now formed a respectable clump and flowers quite freely. Its high-altitude natural habitat should mean that it is the hardiest species of the genus, but this remains to be tested.

Crocosmia pottsii (McNab ex Baker) N. E. Brown
PLATE 11

N. E. Brown, Transactions of the Royal Society of South Africa 20: 264 (1932). De Vos, Journal of South African Botany 50: 486 (1984).

Montbretia pottsii McNab ex Baker, The Gardeners' Chronicle 8: 424 (1877a). *Tritonia pottsii* (McNab ex Baker) Baker, Curtis's Botanical Magazine 109: under plate 6722 (1883). Type: South Africa, Kwa-

Zulu-Natal, without precise locality or collector, cultivated in the garden of horticulturist Max Leichtlin in Baden-Baden, Germany (K, holotype).

pottsii, named in honor of George Honington Potts of Fettes Mount, Lasswade, Scotland, who evidently first grew *Crocosmia pottsii* and provided plants that were grown at the Royal Botanic Garden, Edinburgh, and at Baden-Baden, Germany, in the garden of horticulturist Max Leichtlin, who then distributed it widely.

Plants 28–40 inches (70–100 cm) high. Corm subglobose, about 1 inch (25 mm) in diameter, the tunics papery, light brown, becoming fibrous with age. Stem with two to several well-developed, suberect, weakly diverging branches. Leaves sword-shaped, reaching to about the base of the spike, as wide as $^9/_{16}$ inch (15 mm), medium-textured, plane with a prominent median vein, often with a pair of secondary veins about midway between the median vein and margins. Spike arching outward 45–60°, bearing as many as 30 flowers in two rows; bracts green below, brown at the tips, becoming entirely brown as the flowers fade, $^1/_8$–$^5/_{16}$ inch (4–8 mm)

Crocosmia pottsii, with weaverbird nest hanging from branch above

long, the outer acute or acuminate, the inner with two short teeth. Flowers zygomorphic, nodding, funnel-shaped, orange-scarlet, often paler on the underside, the throat yellow, unscented; perianth tube obliquely funnel-shaped, $9/16$–$13/16$ inch (14–20 mm) long, the narrow lower part about $1/4$ inch (7 mm) long, fairly abruptly expanded; tepals unequal, directed forward, subpatent toward the tips, the dorsal largest, hooded over the stamens, $7/16$–$9/16$ inch (11–15 mm) long, the lateral and lower tepals $5/16$–$1/2$ inch (8–12 mm) long. Filaments unilateral, arching under the dorsal tepal, $3/8$–$1/2$ inch (10–12) mm long; anthers parallel, $1/4$ inch (6–7 mm) long, yellow, pollen yellow. Style arching over the stamens, dividing opposite the anther tips or as much as $1/16$ inch (2 mm) beyond them, the branches diverging, $1/16$–$1/8$ inch (2–3 mm) long, notched apically. Capsules ovoid to globose, broadly three-lobed, $1/4$–$5/16$ inch (7–8 mm) long; seeds angular-prismatic, as many as 8 per locule, 24 per capsule, light to mid-brown, with a spongy coat. Chromosome number $2n = 22$. Flowering time mainly early summer to midsummer (mid-December and January).

The range of *Crocosmia pottsii* is restricted to eastern South Africa and extends from near Umtata in Eastern Cape province through KwaZulu-Natal as far north as Melmoth in Zululand. It is a streamside plant and may be found growing in light bush along the banks of rivers and small streams at elevations of 1000–4000 feet (300–1200 m). Although the individual flowers are relatively small, there are always several open at one time and together make a pleasant sight, the slender stems arching over water. Although no observations of the pollination of the species have been recorded, the flowers are typical of those pollinated by anthophorine bees and worker honeybees. The capsules contain as many as 24 seeds, a feature evidently not before recorded, and they are angular and more or less wedge-shaped and neatly packed together in the capsule chambers. Unlike other *Crocosmia* species, the seeds have a light spongy coat and float on water; most likely they are adapted for dispersal by water.

Crocosmia pottsii may be distinguished from other species of the genus by the small flowers, with a perianth tube $9/16$–$13/16$ inch (14–20 mm) long, and a dorsal tepal as long as $9/16$ inch (15 mm). The flower is funnel-shaped with the tube strongly arched and the tepals directed forward and only weakly curving outward toward the tips. The flowering stem is often several-branched, and individual spikes are fairly lax and have small, dry bracts as long as $^5/_{16}$ inch (8 mm). The leaves are of the ancestral type for the genus: sword-shaped and plane, with a prominent central vein. Perhaps equally unusual in the genus are the capsules, which contain many small, angular seeds with a spongy coat; other species of *Crocosmia* have capsules containing a few large seeds with a hard shiny coat.

Despite their floral and seed differences, we think *Crocosmia pottsii* is most closely related to *C. aurea*. They are the only two species of the genus in eastern southern Africa with plane leaves that have a prominent central vein. Their different habitats as well as flowers and fruits indicate adaptations for different pollinators, seed dispersal agents and habitat preferences. The forest-dwelling *C. aurea* is pollinated by large butterflies, and the large, dark-colored seeds, resembling berries, must be eaten and dispersed by birds. This contrasts with bee pollination, which we believe is ancestral in the genus, and seed dispersal by water in *C. pottsii*.

When *Crocosmia pottsii* first came to the attention of European botanists it was referred to the genus *Montbretia*, a genus now included in *Tritonia*, an African genus of some 28 species (de Vos 1982a, 1983), where it is treated as a section. When *Montbretia* was first placed in synonymy under *Tritonia*, *C. pottsii* was also referred to that genus. It was only in 1932 that N. E. Brown expanded *Crocosmia*, until then including only *C. aurea*. Among the *Tritonia* species he transferred to *Crocosmia* was *C. pottsii*. The confusion about the discovery and naming of *C. pottsii* is related in the chapter, Early Exploration and the Discovery, and need be only briefly repeated. Plants grown in Scotland in the early 1870s by George Honington Potts were passed on to Royal Botanic Garden, Edinburgh, where the gardener, James McNab, recorded it in an unpublished list of outdoor plants grown there as '*Gladiolus pottsii*'. Max Leichtlin of Baden-Baden, Germany, also obtained plants from Potts and, realizing that this was a novelty, sent pressed specimens to J. G. Baker, the bulbous-plant specialist, especially of African Iridaceae, at the Royal Botanic Gardens, Kew. In correspondence preserved at the Kew herbarium, Leichtlin requested that Baker name the species after Mr Potts, which he did in 1877. There is no record of how *C. pottsii* reached Scotland.

Plants of *Crocosmia pottsii* were also obtained at about this time, and most likely from the same source, by the nursery of Victor Lemoine in

Nancy, France. There, *C. aurea* was already in cultivation, and the famous cross between it and *C. pottsii* was made in the late 1870s. Seeds resulting from that cross first flowered in the summer of 1881, and the resulting progeny was subsequently called '*Montbretia crocosmaeflora*', now of course *C. ×crocosmiiflora*.

Hybrid Crocosmias

Here we list plants mistakenly described as wild species or named as if they were wild plants but that are now known to be interspecific hybrids.

Crocosmia ×crocosmiiflora (Lemoine) N. E. Brown, as '*C. crocosmiflora*'

PLATE 12

> N. E. Brown, Transactions of the Royal Society of South Africa 20: 264 (1932). De Vos, Journal of South African Botany 50: 497 (1984).

> *Montbretia ×crocosmiiflora* Lemoine, The Garden 18: 188 (21 August 1881), as '*M. crocosmiaeflora*'. *Tritonia ×crocosmiiflora* (Lemoine) Nicholson, Illustrated Dictionary of Gardening 4: 94 (1888), as '*T. crocosmiflora*'. Type: illustration in C. J. E. Morren, La Belgique Horticole 31: plate 14 (1881), neotype designated here.

> *crocosmiiflora*, from *Crocosmia* and the Latin *florum*, flower, thus 'with flowers resembling those of *Crocosmia*'.

Plants 14–35 inches (35–60(–90) cm) high. Leaves plane, narrowly lanceolate, about two-thirds as long as the stem, $^5/_{16}$–$^9/_{16}$ inch (8–15 mm) wide. Flowering stem laxly branched, the branches ascending and arching outward. Spike lightly flexuose, mostly 6- to 10-flowered; bracts reddish with dry brown tips, $^1/_4$–$^3/_8$ inch (6–10 mm) long. Flowers zygomorphic with a radially symmetric perianth but unilateral stamens, bright orange, paler in the throat, unscented; perianth tube funnel-shaped, slightly curved forward, $^1/_2$ inch (12 mm) long, about $^3/_{16}$ inch (5 mm) wide at the mouth; tepals subequal, mostly $^9/_{16}$–$^7/_8$ inch (15–22 mm) long. Filaments $^9/_{16}$–$^7/_8$ inch (15–22 mm) long; anthers $^1/_4$–$^5/_{16}$ inch (6–8 mm) long, deep yellow. Style arching over the stamens, dividing well beyond the anther tips, the branches about $^1/_8$ inch (4 mm) long, the tips shortly forked. Capsules globose, showing the outline of the seeds, often with aborted seeds, or the seeds globose, reddish brown.

The widely grown *Crocosmia* ×*crocosmiiflora* is remarkably vigorous, disease resistant and easily propagated. It is a primary hybrid between *C. aurea* and *C. pottsii*, made in 1879 in Nancy, France, at the nursery of Victor Lemoine. The flowers are perfectly intermediate between the two species and have the best features of both. The larger flower, with outspread tepals and the bright color, are features *C. aurea* while the multiple branching, numerous flowers, vigor and rapid vegetative reproduction are characteristic of *C. pottsii*.

Crocosmia ×*crocosmiiflora* immediately found favor in western European horticulture. Surprisingly, it remains widely cultivated today because it is so easy to grow, is undemanding of care and soil, and thrives in a wide range of habitats and climate zones. It has long been superseded in the

Crocosmia ×*crocosmiiflora*

nursery trade by larger-flowered or more colorful cultivars made at first by crossing hybrid plants with each other or with the parents, and later with other wild species (see the chapter, Evolution and Classification, under Hybrid Crocosmias). Because of its vigor and rapid vegetative reproduction it persists in abandoned gardens and slowly spreads into meadows, roadsides and eventually into undisturbed vegetation. It is a serious weed in many parts of the world, including Hawaii, North and Central America, Madagascar, Australia, New Zealand, the Philippines and tropical Asia in general. Its distribution across the Indonesian Archipelago is discussed by Veldkamp (1997).

There has been much confusion over the correct author attribution for *Montbretia crocosmiiflora* as well as to the spelling of the specific epithet, sometimes written '*crocosmiaeflora*', '*crocosmiflora*' or '*crocosmaeflora*'. The latter are incorrect compounds according to standard practice in forming Latin specific epithets of this kind. The spelling of the name should simply be corrected to *crocosmiiflora*. The authorship of the name (as a species of *Montbretia*) has been credited to the Belgian botanist Charles Jacques Édouard Morren, who wrote one of the earliest articles about the plant in late 1881, as well as to the editors of *The Floral Magazine,* Frederick William Burbidge and Richard Dean, or the editor of *The Garden,* William Robinson. This is because articles dealing with the plant also appeared in those publications in the summer of 1881 (Kostelijk 1984, Wijnands 1986) when the hybrids first flowered. Perhaps the earliest mention of the plant was in *The Garden* (in August) and *The Gardeners' Chronicle* (July and August). Charles Nelson (1993) argued that a diagnosis adequate to fulfill the requirements of valid publication of the name for the plant may be attributed to Victor Lemoine alone. Lemoine's words were those used in these various publications, and to credit their editors or publishers with formal authorship of the species seems unnecessary. Moreover, it was certainly not the editors' intention to describe a new species. Thus we follow Nelson's proposal to treat '*Montbretia crocosmiaeflora*' as authored solely by Lemoine.

What is to be considered the type of the species is uncertain. De Vos (1984) cited the fine painting in *La Belgique Horticole* as a lectotype, based on the presumption that the author of the article, Morren, also named the plant. Nelson (1993), in his thorough analysis of the problem, suggests that a neotype be chosen since no specimen or illustration can strictly be associated with the earlier descriptions in the British horticultural press. The illustration in *La Belgique Horticole* was based on plants raised by Lemoine, and we designate it the neotype.

Crocosmia ×*crocosmoides* (Leichtlin ex J. N. Gerard) Goldblatt, new combination

Antholyza ×*crocosmoides* Leichtlin ex J. N. Gerard, Garden and Forest 10: 315–316 (1897), as '*A. crocosmoides*'. Type: of cultivated origin, released commercially in the Netherlands by the van Tubergen bulb company in 1895, and herbarium specimens sent to the Royal Botanic Gardens, Kew, in 1904 (K, neotype).

Crocosmia ×*latifolia* N. E. Brown, Transactions of the Royal Society of South Africa 20: 264 (1932), as '*C. latifolia*'. De Vos, Journal of South African Botany 50: 483 (1984). Type: of cultivated origin, released commercially in the Netherlands by the van Tubergen bulb company in 1895, and herbarium specimens sent to the Royal Botanic Gardens, Kew, in 1904 (K, syntypes).

crocosmoides, from *Crocosmia* and the Greek suffix -*oides,* -like.

Plants 20–40 inches (50–100 cm) high. Leaves lanceolate, lightly pleated and with multiple major veins, about two-thirds as long as the stem, $^{13}/_{16}$–$1^9/_{16}$ inches (2–4 cm) wide. Flowering stem laxly branched, the branches ascending and arching outward. Spike flexuose, with many flowers; bracts reddish brown with dry brown tips, $^3/_8$–$^9/_{16}$ inch (10–15 mm) long. Flowers zygomorphic, orange-red, probably unscented; perianth tube trumpet-shaped, slightly curved forward, $^9/_{16}$–1 inch (15–25 mm) long, widening gradually from the base to the mouth, about $^3/_{16}$ inch (5 mm) wide at the mouth; tepals unequal, the dorsal largest, as long as $^7/_8$ inch (22 mm), lateral and lower tepals $^1/_2$–$^{11}/_{16}$ inch (12–18 mm) long. Filaments $^{11}/_{16}$–$^{13}/_{16}$ inch (17–20 mm) long; anthers $^1/_4$–$^5/_{16}$ inch (7–8 mm) long, yellow. Style arching over the stamens, dividing opposite the anthers, the branches about $^3/_{16}$ inch (5 mm) long, the tips shortly forked. Capsules and seeds unknown.

Crocosmia ×*crocosmoides* is the result of a deliberate cross made by Max Leichtlin of Baden-Baden between two species, *C. aurea* and *C. paniculata,* then known as *Antholyza,* the latter a plant he had recently acquired.

The hybrid was called *A. crocosmoides,* evidently by Leichtlin, who made his new hybrid available about 1890. The name seems first to have appeared in print, without description, in the van Tubergen catalogue of 1895. Then, in August 1904, preserved specimens were sent by van Tubergen to the herbarium at the Royal Botanic Gardens, Kew; the appearance of these specimens confirms the parentage. The plants have leaves only lightly pleated but with multiple veins toward the base, rather than a single prominent midrib, which indicates one parent species with pleated leaves and the other with plane leaves. At the time, that is, before 1904, *C. paniculata* was the only pleated-leaved species of *Crocosmia* in cultivation in Europe, and the shape of the flowers of the hybrid makes it all but certain that *C. aurea* was the other parent.

'*Antholyza crocosmoides*' remained a botanically invalid name for some years because it lacked a formal description or even a few words describing its important features in contemporary catalogues. In 1897, however, John N. Gerard (fl. 1888–1897) described *A. crocosmoides* in some detail, also noting its cultural requirements and parentage (Gerard 1897). Botanically this is regarded as sufficient for valid publication, and Gerard must be regarded as the author of the name, albeit an inadvertent one. The plant was treated in Liberty Hyde Bailey's (1914) *Standard Cyclopedia of Horticulture,* there spelled '*crocosmioides*', but otherwise seems to have been largely ignored in botanical or horticultural literature. The name was never listed in indexes to botanical names, presumably because of its hybrid origin. (The history of *C. ×crocosmoides* in cultivation is discussed in more detail in the chapter, Cultivation.)

Then in 1932, in his expanded treatment of *Crocosmia,* Nicholas Edward Brown described *C. latifolia,* evidently accepting the plant as a naturally occurring species. The name was based solely on the specimens preserved in the herbarium at Kew, sent there as *Antholyza crocosmoides* by van Tubergen in 1904 and bearing that name, a fact Brown failed to note. *Crocosmia latifolia* was accepted by Miriam de Vos in her 1984 revision of the genus, but with reservation because it was unknown in the wild. Later, in her 1999 account of *Crocosmia* for the *Flora of Southern Africa,* de Vos removed it to the end of her enumeration of the species, placing it together with *C. ×crocosmiiflora* under the heading, Hybrids. It is clear that *C. ×latifolia* is a later name for *A. ×crocosmoides* and so should be placed in synonymy. We thus make the new combination *C. ×crocos-moides* for the plant, preserving the original, seemingly incorrect spelling as required by the *International Code of Botanical Nomenclature*. It may seem odd to take the trouble to make a new combination for a hybrid plant that is today at best rare in cultivation. In defense, we note that a name for this hybrid is needed because it is the parent of several cultivars, some still grown. Surprisingly, *C. ×crocosmoides* remains in cultivation at Kew. The plants there are almost certainly the very ones that were obtained from the van Tubergen bulb company in 1904 and are thus more than a century old. They will probably continue to thrive until deliberately removed from their comfortable quarters.

Cultivation

Crocosmia and *Chasmanthe* are normally deciduous plants that have an extended dormant period in the dry season, the summer months for *Chasmanthe* and *Crocosmia fucata,* or the winter months for other *Crocosmia* species. New plants grow from corms, which act as food reservoirs for the plant, as well as dormant growing tips. *Crocosmia paniculata* can reportedly be evergreen in the wild, but this is not a normal characteristic of the species, and in cultivation, even in exceptionally mild, frost-free winters, it is deciduous. In areas with very warm winters, however, the *C. ×crocosmiiflora* hybrids tend to remain evergreen. Lack of a period of dormancy weakens the following year's growth, and in the warm climate of Zimbabwe in southern tropical Africa, commercial growers of crocosmias for the cut-flower trade lift the corms in at the end of the growing season and keep them cold to provide this dormant period.

Chasmanthe

The three species of *Chasmanthe* all grow in similar conditions in their native habitat, the winter-rainfall belt of southern Africa at the southwestern corner of the subcontinent. Requiring similar cultural conditions, they are readily grown in relatively fertile and freely draining compost. Watering is only required when they break summer dormancy, and their growth begins in early autumn. They will survive being kept moist in summer, providing the soil is well drained, but if possible they should be left as dry as conditions allow. The corms increase readily, so frequent division or repotting into larger pots is required. New corms are made on top of the old ones, which are not normally viable and will shrink or decay with age. The primary corm, which has a small growing shoot or bud at its apex, will grow out, producing a fan of leaves and, normally, a flowering stalk. While the small lateral corms have a typical 'bulb' shape, the mature corms are distinctively flattened and sometimes as much as 4 inches (10 cm) in diameter, but barely $^3/_8$ inch (1 cm) thick. Disturbance of the corms can delay flowering by a year, and application of a weak liquid feed will usually provide better results when the corms have not been re-planted.

Depending on the conditions of cultivation and the temperature under which *Chasmanthe* species are grown, flowering can occur in the Northern Hemisphere between November and February in Europe and California. *Chasmanthe aethiopica* flowers earliest, always before December. It is followed by *C. bicolor* in December and January and is usually overlapped by *C. floribunda,* which will reach peak flowering in January or February. In the Southern Hemisphere, the pattern follows closely that of the wild species; thus *C. aethiopica* will flower soon after the first winter rains (or artificial watering) in May and June, sometimes into July. *Chasmanthe bicolor* follows the same pattern, but usually flowering first in June and lasting until early August. Last is *C. floribunda,* which rarely flowers before July and may be found in some sites in South Africa still blooming in September.

All three *Chasmanthe* species were grown in Europe in the early 19th century, two of them even earlier. *Chasmanthe floribunda* was actually grown in Paris as early 1633. Adapted to cool (but normally cold) wet winters and dry summers, plants were always grown under glass in houses heated by coal fires. The dry heat suited these Cape plants, and their winter and early spring flowering made them popular conservatory or winter garden plants in the 19th century. By the 1920s or 1930s the age of the stove house was over, and *Chasmanthe,* not normally hardy in northern Europe, largely disappeared from cultivation there. The corms of *Chasmanthe,* like many bulbs of Mediterranean climate zones, require a period of dry baking in their dormant period and tend to rot in ground that remains cool and moist in summer.

There is no evidence that Victorian nurserymen ever attempted to hybridize *Chasmanthe* despite their tendency to try to hybridize most plants newly brought into cultivation. Possibly, the similarity of the flowers of the three species in the genus deterred any attempt at hybridization as likely to be unproductive. The only record of any deliberate hybridization in *Chasmanthe* is the relatively recent crossing of the two color forms of *C. floribunda* in Australia by David A. Cooke, reported on his Web site in 1998, 'Flower colour in *Chasmanthe floribunda*'. He concluded a single gene is involved in the difference in flower color, and the yellow form, often called *C. floribunda* var. *duckittii*, simply lacks the gene for orange pigmentation.

As a result of the decline of large conservatories, *Chasmanthe* species have lost whatever popularity they had in Great Britain and northern Europe, yet they remain in cultivation in gardens in climates more suited to their needs, often growing in south-facing beds adjacent to the base walls of heated conservatories in botanic gardens, which keeps the soil warmer in winter than would otherwise be the case. Both color variants of *C. floribunda* are excellent plants for the garden in areas of mild winters, especially those with Mediterranean climates, such as southern Europe, California, central Chile and South and Western Australia. Indeed, *C. floribunda* has become naturalized locally in California and is almost weedy in places (Goldblatt 2002) as is also the case in many old gardens of the French Riviera, southern Italy and even the famous gardens of Tresco Abbey in the Scilly Isles off the coast of Cornwall, England. They also thrive in the somewhat frosty, dry winters of the South African high veld, where they must be watered during the growing season. The plants have handsome foliage and when in bloom make a striking display for 3–4 weeks in spring; the upright, elegant foliage is attractive for another month while the fruits develop, carried in the orange form on elegant bronze-red stalks. Plants will die back in summer, even in places where there is ample summer rain. They can be interplanted with a summer-blooming bulb so the ground is not left bare while they are dormant. The winter-deciduous, pale-blue-flowered *Agapanthus campanulatus* or summer-flowering *Watsonia pillansii* make ideal companions for *C. floribunda* in the garden.

The last of the *Chasmanthe* species to be named, *C. bicolor,* seems again to have been first grown in France. In John Lindley's 1828 article in *The Botanical Register,* in which he named it *Antholyza aethiopica* var. *minor,* he noted that the painting was obtained in the garden of the Comte de

Vandes 3–4 years before. Records from Vienna are even earlier, and specimens preserved there in the herbarium of the natural history museum bear the date 1811. It was soon dispersed to gardens elsewhere in Europe and was actually described in 1832 from plants grown in Sicily. The plant was named *A. bicolor* by Guglielmo Gasparrini, based on plants grown near Palermo in the no longer extant botanic garden at Boccadifalco. Just 100 years later, the yellow variant of *C. floribunda,* called variety *duckittii,* was described in South Africa by Louisa Bolus.

Crocosmia

The introduction into cultivation of the species from this small, mostly attractive genus has spanned two centuries. It is perhaps surprising that *Crocosmia fucata,* the first species to be cultivated, is one of the rarest in the wild. Despite being apparently quite hardy, it has only relatively recently been reintroduced into cultivation in Europe and probably has not been cultivated, even in South Africa, for well over 100 years. It is to be hoped that it will not again take almost 25 years to flower in cultivation, as was the case after its original introduction. Its shyness to flower may have been the result of its being grown in too dry conditions. Similar shyness when dry is a characteristic also shown by *C. pottsii,* which like *C. fucata* favors wet conditions in the wild. In contrast to *Chasmanthe,* species of *Crocosmia* have a wide range across Africa south of the equator, and the several species also grow in a range of habitats. The only winter-growing species of *Crocosmia* is *C. fucata,* from the mountains of Namaqualand in western South Africa. It was grown successfully in Yorkshire, England, where it flowered in 1838. In the wild, it flowers from late spring (late October) until early summer (early December).

    Crocosmia fucata, C. mathewsiana, C. paniculata and *C. pottsii* are all found in relatively damp habitats, and except for *C. pottsii* at quite high altitudes, so they should thus be the hardiest crocosmias. *Crocosmia masoniorum* and *C. pearsei,* also from relatively high elevations, occur in well-drained habitats, quite dry in the winter months. With its wide geographic range, *C. aurea* occurs at low elevations in coastal forest, and plants from such places are the least hardy of the genus. *Crocosmia aurea* also grows in mountain forest at the same elevation as *C. mathewsiana,* and plants from higher altitudes may prove more hardy than those already in cultivation.

While *Crocosmia fucata*, *C. mathewsiana*, *C. paniculata* and *C. pottsii* are reasonably hardy, as a general rule it is best to assume that no *Crocosmia* or *Chasmanthe* corm will withstand being frozen, so corms should be planted at a depth below that which frost will penetrate. If frost normally penetrates to a level of about 6 inches (15 cm), a mulch should be applied. In areas where the ground is regularly frozen to a greater depth, corms should be lifted in winter and stored in frost-free conditions. It is best if they are potted up and kept slightly moist, as those of most species do not tolerate being dried out. Crocosmias are often sold as dry corms, and these are best potted up in cold conditions so that they overcome their desiccated state before growth begins. Substantial losses may result if dry corms are planted in warm moist conditions.

In *The English Flower Garden* (Robinson 1933) one technique for the winter care of crocosmias is perhaps worth quoting. The cultural directions were written in 1901 by Rev. Charles Wolley-Dod, who gardened in Cheshire, in north-central England. He was writing primarily about the then new cultivars of *Crocosmia ×crocosmiiflora* at a time when the winters were much more severe than they are today. The directions provide an interesting insight into an earlier age of gardening:

> To make them do well they must be kept thin, and so must be divided every year. This may be done at any time of year before the ground is frozen up. My practice at Edge, after digging them up—suppose there are twelve stalks, that is twelve bulbs in each clump, with three or four young points to each bulb—is to have fifty or a hundred pots ready and put three bulbs into each pot, filling up with any waste soil, drainage being superfluous. The less they grow before March the better. They must not be cut down till spring. When all the pots are full they are placed together in some sheltered spot out of doors and well watered— for if kept dry they die—then they are covered with a foot or two [30–60 cm], according to the weather, of dry leaves or other litter, enough to ensure their safety from frosts. By the end of March they are safe, and may then be planted out anywhere, letting the bulbs at least six inches [15 cm] deep, either amongst

herbaceous plants which they like or amongst low shrubs. I have some in beds of dwarf roses where they do look very well. As they increase at least fourfold every year, the gardener must harden his heart and not be tempted to let them grow more densely; but as he will find most of his friends have as many as they want, throw the surplus on the rubbish heap. I find one morning each year enough for this work, which may be done in roughest and most hasty way without detriment to the bulbs. Indeed I have sometimes buried the clumps in a soil heap, for the winter, littering them over as described, and planting the bulbs out by threes in spring. The main objects are not to let them get frozen and not to let them get dry and grow in winter. I generally also replant three bulbs where I dig up each clump. If the winter is mild these survive and the pots are not wanted; if they are killed the pots take their place. They flower better if a spadeful of rich stuff is put in where each pot is planted.

The head gardener at Westwick Hall in Norfolk and the first significant breeder of crocosmias in England, George Davison (1905) demonstrated a greater understanding of the conditions which crocosmias required:

> Locality and soil possibly make a difference. It is sometimes recommended that the bulbs be taken up and dried, which I consider very injurious to them. A friend of mine read an article in a paper which advised to take up the plants in October, divide and replant them. He followed this advice with disastrous results, losing every one in the winter, wet and frost no doubt rotting the bulbs. My friend weakened his plants by taking them up, as they could not get a root-hold again before winter. They will survive the winter in any light soil, if left undisturbed. In January 1895, we registered frosts below zero [0°F or −20°C]; my montbretia bulbs were destroyed but the stolons were alive, and I planted them in March and had a fine border of flowers.

In short, *Crocosmia* corms should be kept cool, moist and frost free in winter, but there are many easier ways to achieve this than those described by Wolley-Dod. The corms resent being dried out, frozen or being kept in very cold, wet ground in winter. Annual division is not necessary; replanting every 2–3 years is necessary for the more vigorous *C. ×crocosmiiflora* cultivars, but many cultivars of other pedigree and most species may be left undisturbed for rather longer. Some very vigorous hybrids may need almost annual thinning and replanting or feeding, but they are usually the less garden-worthy plants. Thinning and replanting are best carried out in early autumn or spring, when the soil is warm enough to allow the plants to reestablish quickly.

In cultivation, all crocosmias, both species and hybrids, will grow well in the classic ideal soil type: a reasonably fertile, friable, moisture-retentive but freely draining soil. Many of the original breeders and growers of the early hybrids regarded crocosmias as gross feeders, but excessive fertilization is not an absolute requirement. They will perform quite adequately without being overfed. The flowers of the larger-flowered cultivars will be a little larger if grown richly, and regularly divided and the soil improved when replanted. Incorporation of well-rotted compost into the soil, particularly if it is fairly poor or starved, or a light dressing of bonemeal or granular fertilizer will suffice to ensure a good display provided there is adequate moisture. Alternatively, feeding with liquid tomato fertilizer should prove beneficial. High-phosphate liquid feeds, currently marketed for the saturation of annual bedding plants, may produce some improvement in the performance of crocosmias, but such overfeeding in not necessary.

There is perhaps one exception in which crocosmias may benefit from very fertile growing conditions. The better cultivars, if starved and grown in poor conditions, will become severely weakened, producing smaller than normal corms. They require several years of cultivation in good conditions to recover and show their true potential. Such corms would probably do best with a high-nitrogen but low-phosphate fertilizer initially, to initially encourage foliage but discourage flowering until the corms fully recover.

In general, the period of dormancy for any *Crocosmia* will vary with the severity of the winter. It can be quite brief in a mild winter. In a mild autumn, new shoots may start to appear even before the foliage of the current year's growth has fully died down. This is particularly evident in *C. pottsii* and some forms of *C. ×crocosmiiflora*. Subsequent frosts can damage young exposed foliage, but usually without any long-term effect on the plants, as George Davison experienced. Even before, Wolley-Dod (1901) reported that if young shoots were frozen hard down to the corm, they would not regrow. In other words, the corms cannot withstand being frozen. While there is some variation in the hardiness of the corms of different species and cultivars, it is safest to assume that none is frost hardy. *Crocosmia aurea* is without doubt the least hardy. In marginal conditions there is an increased likelihood that corms will survive if the soil is light and freely draining or relatively dry.

Growth habits will vary in different climatic regions, and the comments here are based on the relatively cool maritime climate of the British Isles and the Atlantic coast of France but should apply equally to the Pacific Northwest of North America, southern and eastern China, Japan, New Zealand, and at least eastern Australia and eastern southern Africa. In a normal season, plants can be expected to die back in mid- to late autumn (October–November in the Northern Hemisphere), hastened by the first frosts. Different cultivars and species die back at different times. *Crocosmia aurea* and closely related cultivars are usually the first to start dying back despite being the last to flower. They can even die back without flowering if the summer has not been hot enough or long enough. *Crocosmia masoniorum* and its forms and hybrids can be a little variable, while *C. mathewsiana, C. paniculata, C. pearsei* and *C. pottsii* seem to be more resilient to the onset of winter. This is not surprising as they come from higher elevations and are generally hardier. *Crocosmia ×crocosmoides,* showing the influence of *C. paniculata,* is also quite late in dying back.

New growth, generally in spring, is related to soil temperature. Mild spells in winter can also stimulate premature new growth. The main growing activity usually starts in early or mid-spring (March or April in the Northern Hemisphere) depending on the season. Limited new foliage growth, to about 6 inches (15 cm), can occur in cool springs, with further development deferred until there is a significant increase in temperatures. This suspended growth can last in excess of 2 months, or until soil temperatures have risen to about 22°F (12°C) or more.

In the wild, most crocosmias are found in areas of summer rainfall and cold, dry winters. They do, however, seem remarkably well adapted to the mild, moist winters of the British Isles and to those of the Southeast and

Pacific Northwest of North America, if not to the extremes of cold. With the exception of *Crocosmia pottsii,* and very likely *C. fucata,* which grow in relatively damp conditions, the species and most cultivars will not grow as well in cold and wet heavy or clay soils. It is likely that the combination of winter wet and frost does the most damage and can cause corms to rot or freeze.

Various species have different tolerances to cold and moisture. *Crocosmia pottsii* is considered to be the hardiest; it is certainly the best adapted to wet conditions in winter, considering its natural habitat. *Crocosmia masoniorum, C. paniculata* and, presumably, *C. pearsei* should also be very tolerant of cold. In the wild, they are usually found at reasonably high elevations, growing on fairly freely draining stony ground; it is perhaps more doubtful if they are tolerant of very wet conditions, particularly if combined with cold; *C. masoniorum* is almost certainly susceptible to excessive damp (Bloom 1965). It is likely that *C. pearsei* will not tolerate excessive damp either. *Crocosmia mathewsiana* seems to be reasonably tolerant of damp conditions although in the wild it grows in sandy, well-drained ground but in a relatively moist climate.

Cultivars of *Crocosmia aurea* are generally considered to be the least hardy, though one would expect that the wide geographical range of the species would produce variation in the climatic tolerance of the plants. Although all the other species are usually found in the open in full sun, *C. aurea* is native to shady habitats, usually shade in evergreen forest or in forest gaps and margins. Wolley-Dod asserted that *Crocosmia* cultivars do best rooted in cool moist conditions but in the sun, circumstances that favor quite a number of plants. Cooler ground conditions certainly delay the flowering of crocosmias, but adequate moisture is necessary for good flowering of most. *Crocosmia pearsei* may be the exception as it seems to prefer hot, dry conditions, as does *C. masoniorum* to a lesser degree. Neither species fares well in rich moist soil. *Crocosmia pearsei* is exceptionally slow to increase, though seems to increase better in hot, dry conditions. *Crocosmia masoniorum* increases more readily but probably does better with slightly more moisture.

Corms of most crocosmias do not thrive if dried out, especially *Crocosmia pottsii* as it is normally continuously exposed to high levels of moisture. Corms of *C. ×crocosmiiflora* and *C. masoniorum,* and some of the better hybrids can survive such desiccating treatment, otherwise gardeners and nurseries would not be able to grow the dry, mass-market *Crocosmia* corms. In the Netherlands, crocosmias have to be lifted in autumn so that the corms are not killed by penetrating winter frosts. They are normally stored dry to prevent mold-induced corm rot. Corms are usually replanted in early spring once the risk of deeply penetrating frosts has gone but when the soil is still cool enough to allow the corms to rehydrate before starting into growth. Planting in too warm conditions may be the cause of a relatively high failure rate of such dry corms being potted up in the nursery trade. Flowering can be delayed beyond the normal flowering season by delaying planting, and growth and flowering can be put into suspended animation in cold storage at about 39–41°F (4–5°C). These are among the many techniques used by exhibitors at flower shows to produce flowers at their best on a particular day or even out of season.

In climatic regions with severe and penetrating winter frosts, *Crocosmia* corms should be lifted in autumn and stored in frost-free conditions. In intermediate regions a heavy mulch may suffice to provide winter protection. Alternatively, crocosmias may be treated as pot plants rather than planted in the ground, with the pots stored in frost-free conditions in winter. Repotting each year would be advisable, with a thinning out of the corms if necessary. The advice on winter care of crocosmias by Wolley-Dod, republished by William Robinson, suggests an extreme level of care and treatment that only a bygone age of gardening could sustain.

As noted, *Crocosmia masoniorum* and *C. pearsei* will withstand and may do better under drier conditions. *Crocosmia paniculata* will survive dry conditions but will become shy to flower. *Crocosmia aurea* and its hybrids, and the better *C. ×crocosmiiflora* cultivars, require fairly rich moist conditions with relatively high temperatures to give their best display. Size and quality of the flowers can vary considerably, depending on how well they are grown.

Most *Crocosmia* hybrids increase by stolons, which start to develop in autumn and produce fresh corms during the new season's growth. Corms develop simultaneously with the new growth. This characteristic is inherited from both *C. aurea* and *C. pottsii*. Corms of *C. aurea* each produce several stolons that send up new growth points as far away as 2 inches (10 cm) from the parent corm. *Crocosmia aurea* subsp. *pauciflora,* a small plant with the smallest corm, also produces the finest and shortest stolons. Each new growth point has the potential to flower and, in ideal growing conditions, frequently does.

Crocosmia aurea requires heat as well as moisture to grow well and flower. It grows in fairly deep shade in its natural habitat and will do so in cultivation. It will also grow well in full sun in more northerly climates. In contrast to the other species, it does not initially make a compact clump but sends out several stolons from which several new growth points emerge, 2 inches (5 cm) or more away from the parent corm. In time, density will increase, resulting in the formation of a substantial clump, by which time the corms will require some feeding or thinning. *Crocosmia aurea* is the only species that does not naturally build up a chain of new corms directly on top of the old ones. We have seen it in the wild growing in evergreen native forest in almost complete shade together with shade-loving *Aristea ecklonii* and *Dietes iridioides,* all surrounded by *Plectranthus.* This shade tolerance may be exploited in warm gardens to advantage.

In good growing conditions, *Crocosmia pottsii* increases much too vigorously to be welcome in many gardens, except in larger ones or semiwild areas. It will quickly become shy to flower once it has formed a solid clump, or will not flower at all if grown in conditions that are too dry. The noted plantsman Fredrick W. Burbridge (1890a, b) lamented growing the species for many years but never succeeding in flowering it and sought advice as to the secret of success. Its vigor is a characteristic passed on to *C. ×crocosmiiflora,* its hybrid with *C. aurea* and commonly known as montbretia, and more recently to some hybrids with *C. masoniorum.* Each corm will produce several short stolons, on some of which intermediate corms are formed and from which many more stolons are produced, all of which will produce foliage and new corms the following year. Complete removal of an unwanted clump can be problematical, as any small piece of stolon seems capable of regeneration, the unwelcome property shared by most pernicious weeds. Spraying the foliage with a glyphosate-based weed killer several times during the growing season is the most effective way of killing the plant before attempting to dig up the dead corms. Organic gardeners are faced with leaving the area fallow for probably two growing seasons after digging up the live corms, and frequent removal of all subsequent shoots.

In the wild, *Crocosmia pottsii* can be found growing at the edges and even in streams, though not in such substantial clumps. For some reason its potential as a pond plant has not yet been exploited in cultivation. Yet its small but bright red flowers and later flowering period than most pond plants should make it a welcome addition to many water gardens. Grown in water, its prodigious vigor is significantly reduced. In water it will require planting in a medium that provides some nutriment, otherwise it may become shy to flower. *Crocosmia fucata* may also grow well as a plant for pond margins, though it is likely to be some years before its potential can be assessed.

Crocosmia pearsei, though occurring singly in the wild, will increase vegetatively, but very slowly, in cultivation. It seems to increase more readily in relatively warm, dry conditions, as can be seen at the Royal Botanic Garden, Edinburgh, where it has formed a respectable clump and flowers quite freely in a fairly hot, dry bed near the Alpine House. Its high-elevation natural habitat should mean that it is the hardiest member of the genus. Its slowness to increase and rarity in cultivation are likely to delay testing its hardiness in extreme cold. It is unlikely to tolerate very wet conditions in winter. Dry conditions seem to induce the corms to send out the occasional stolon rather than just make a continuous chain of corms.

Crocosmia masoniorum seems to prefer less-rich and relatively drier conditions to grow well and increase, though it will grow quite well in clay soils. The normal flower color is orange but a rich chrome yellow form is known from the wild, which is perhaps apt for a plant with the common name golden swan crocosmia in South Africa. There is a good yellow ('Rowallane Yellow') and at least two red ('Flamenco' and 'Dixter Flame') cultivars being grown in the British Isles. These color forms cannot be grown alongside one another without getting intermingled because of the stoloniferous growth habit. This is not immediately apparent. New corms are formed on top of old ones, but stolons are also produced and satellite corms formed along the stolons as far away as 2 inches (10 cm) from the parent corm. They do not produce foliage until their second year; by that time another satellite corm will have been formed farther away. Once the corms produce foliage they make an additional attached corm annually and flower in their third or fourth year. If adjacent clumps have begun to intermingle, the only way to rogue the clumps is to separate out corms in flower. In time, substantial clumps are built up with a solid chains of large corms among a mass of smaller satellite corms, which make digging up such clumps difficult. Live foliage and flowering stems are extremely brittle where they attach to the corm and so are easily detached, which makes roguing especially difficult. *Crocosmia masoniorum* can be shy to flower and will normally take a year to settle in after planting before even clumps of corms will start to flower again, though medium-sized to large corms separated out and replanted are more likely to flower in the first year.

Crocosmia mathewsiana increases slowly but remorselessly by branching chains of corms similar to those of *C. paniculata* and seems to flower quite freely and regularly. In cultivation it is most similar to *C. pottsii* in size and flower color, but its open branching habit and graceful drooping spikes make it a more attractive garden plant. Its rareness in cultivation has so far limited the testing of its tolerance to extremes of cold and moisture. Its relatively late flowering in a cool temperate climate, a week or two before *C. aurea,* and well-behaved habit should make it a worthwhile introduction into more widespread cultivation. It seems to be quite tolerant of dry conditions, for although it grows in a warm, humid climate, its native habitat is sandy and well drained.

In cultivation, *Crocosmia paniculata* seems to be the most adaptable species of all, providing conditions are neither too dry in summer nor too wet or cold in winter. It increases by new corms growing from the older ones in a solid chain formation. More than one new corm can grow from an old one, so the chains branch and in time will form quite massive and solid clumps that require considerable effort to remove. Flowers are small in comparison to the size of the plant, but the pleated foliage is an attractive foil for smaller plants grown in front of *C. paniculata.*

The plant commonly and for long in cultivation in Europe as *Crocosmia paniculata* is in fact a hybrid with *C. pottsii.* When this hybrid was raised or by whom is not known, but as it is perhaps more garden-worthy it seems to have largely supplanted the true species, which is now quite rare in cultivation. The hybrid is more freely flowering, has slightly larger, red rather than orange and yellow flowers, and more conspicuously pleated leaves. In time, it forms even more substantial clumps, which require considerable effort to dig up as the large corms are all directly connected to one another. Fortunately, it has not inherited the prodigious power to increase from *C. pottsii* and if anything is slower to increase than the true species. In eastern South Africa, in Mpumalanga and KwaZulu-Natal provinces, a similar, tall plant with pleated leaves and dark red flowers is fairly widely established and though probably not the same plant as that in western Europe, which was introduced into cultivation about 1890, is also confused with true *C. paniculata*. We wonder whether it may be the result of natural hybridization.

The most common crocosmias in cultivation are the hybrids, predominantly *Crocosmia ×crocosmiiflora* and the larger-flowering cultivars descended from this original cross. The vigor, adaptability and tenacity of the original hybrid are testified by its widespread distribution in temperate areas, right around the globe, from the Himalayas to Hawaii in the Northern Hemisphere, and New Zealand and Madagascar in the Southern Hemisphere.

Some of the large-flowered cultivars produced in the heyday of breeding are closer to *Crocosmia aurea* in habit and lack of hardiness, which may account for their failure to survive in cultivation; others are closer to *C. pottsii*. The vigor inherited from this latter parent is another likely reason why so many of the large-flowered cultivars have not survived in cultivation. The rapid increase of the plants results in progressively smaller flowers and eventual nonflowering if the plants are neither fed nor fairly regularly divided. These hybrids send out a varying number of stolons, and once the corms in the center of a clump become crowded, so that they cannot send out stolons, they make new corms on top of the old ones, a characteristic inherited from *C. pottsii*. Thus in old clumps the corms gradually mound up out of the soil. This leaves the topmost exposed corms vulnerable to severe frosts. However, unlike two other widely cultivated genera of the *Iris* family, *Dierama* and *Watsonia,* older corms of *Crocosmia* do not die after the new annual corms are formed on top but remain alive a number of years. Thus old, nonflowering clumps may be reinvigorated by separating out the corms, even from the vertical chains of corms resulting from many years' growth. The vertical chain of corms is a characteristic inherited from *C. pottsii;* many of the old cultivars, in which the genes of *C. aurea* are dominant, do not form such vertical chains of corms. Their small satellite corms, borne on short, wiry stolons, become more crowded and reduced in size. Surprisingly, lower corms will flower potentially better than the top corm, though the new growth is a little later and less foliage is usually produced. They can thus give a better display of flowers.

More recent hybridization between different species, such as *Crocosmia masoniorum* and *C. paniculata,* has resulted in the production of many cultivars that do not multiply as quickly and seem to require less feeding. Large clumps will remain quite floriferous, such as those of *Crocosmia* 'Lucifer' (Plate 13). There are a number of cultivars originating in the Netherlands that are the result of crossing *C. pottsii* with *C. masoniorum.* One or two of these hybrids combine the prodigious vigor of increase of the former with the stoloniferous spread of the latter, resulting in plants that are a liability in any temperate garden and difficult to eradicate. Unfortunately, the worst such cultivar is sold under a variety of names by Dutch suppliers, including *Crocosmia* 'Marcotijn' and as the species *C. masoniorum.* Performance during the first year provides little warning of this undesirable feature.

In the final analysis, the performance of the plants themselves is the ultimate indicator as to their particular needs. When the plants require feeding or division and replanting, they will become congested, shy to flower or shorter and weaker in growth (most probably both). Growing crocosmias in too dry conditions will, however, limit the flowering of most cultivars and species and also produce shorter and weaker growth.

In cultivation, the order of flowering of the species seems to exhibit some correspondence with the altitude and latitude of their natural habitat. *Crocosmia pearsei* is the earliest to flower, followed by *C. masoniorum,* then *C. paniculata* and *C. pottsii,* which flower about the same time. There is some variation in the flowering period between different forms of both these latter species, which may relate to the location or elevation of their natural habitat. *Crocosmia aurea* and *C. mathewsiana* are the last to flower, as they are in the wild. In particularly warm conditions, *C. mathewsiana* apparently flowers much earlier, as might be expected. *Crocosmia ambongensis* and *C. fucata* are not in cultivation, so their positions in this sequence cannot be established, but as *C. fucata* normally flowers in November in its Southern Hemisphere habitat, it is likely to be one of the earliest species to flower, in spring rather than summer in cultivation in the Northern Hemisphere. The species from higher elevations are adapted to flowering, and indeed seeding, at lower temperatures.

The actual month of flowering of crocosmias in cultivation will depend on the climatic conditions where they are grown. The warmer the spring, and hotter the summer, the earlier the plants will flower and the shorter the flowering period. In cooler conditions the flowering period of a range of species and cultivars may stretch out over 4 months; in hot conditions it may barely last 2 months. The initial flowering, in the Northern Hemisphere, of *Crocosmia pearsei* can vary from mid-May to late June, and that of *C. aurea* and *C. mathewsiana* can extend from early September in a warm climate to mid-October in cool conditions. In poor autumn conditions the late-flowering species and cultivars may not get a chance to flower at all.

High temperature also has a pronounced effect on the flowers of *Crocosmia aurea*. The long, narrow tepals react to high temperatures like a bimetallic strip, with the tepals recurving backward like the flowers of *Lilium martagon* or an *Erythronium*. This characteristic seems to be triggered by temperatures of approximately 77°F (25°C). High humidity at lower temperatures may also trigger the reflexing curvature. In the cool of the evening, the tepals straighten again or even close slightly, only to recurve again in the heat of the following day. This characteristic is present in some of the cultivars and is pronounced enough in one for it to be called *Crocosmia* 'Martagon'.

While the appearance of most of the cultivars is fairly constant, given reasonable growing conditions, the appearance of some is variable, depending on temperature. This seems to be most prevalent in cultivars that are closely related to *Crocosmia aurea* and those with the genes of *C. aurea* subsp. *aurea* 'Maculata' (Plate 5). The distinctive brown flecks on the inner three tepals of this naturally occurring form seem to be temperature-related. They are quite faint in cooler conditions, more pronounced in warm conditions, and when well grown in hot conditions the marks can appear on all six tepals.

The largest-flowered cultivar, *Crocosmia* ×*crocosmiiflora* 'Star of the East' (Plate 13), develops one or two dark markings in the center in warmer conditions and occasionally will have a purple tint in the center.

The next largest-flowered, 'His Majesty' (Plate 14), will also develop purple blotches at the base of the several segments in hotter conditions. In warm conditions most of the red pigment in the tepals fades to gold, leaving only the tips tinted; in hotter conditions the tepals seem to retain much more red pigment, leaving only the center gold. Thus 'His Majesty' can produce the most dramatic change in appearance of all. 'Nimbus' (Plate 14) is pale yellow with a red-brown circle of halo around the center that is more pronounced in hot conditions. One of the most widely grown cultivars, 'Emily McKenzie' (Plate 15), has a pale yellow center in hot conditions but an orange center in cooler conditions, similar to the color of the segments beyond the maroon markings.

Another form of color change can occur as the flower matures. Different forms of *Crocosmia* ×*crocosmoides,* the cross between *C. aurea* and *C. paniculata,* exhibit a continual variation in the extent of the red flush on the outside of the orange tepals as the flower opens and matures, with the last few flowers becoming predominantly yellow. This is likely to be accentuated in warmer conditions. Likewise, a similar continuous variation occurs with the extent of the red flush on the inside of the segments of 'Comet' (Plate 14), the hybrid between *C. masoniorum* and *C. aurea,* as the flower matures. The predominant red flush on the inside of the yellow segments can gradually reduce to little more than a red line along the length of the segments, but the predominant red suffusion seems to last longer in warmer conditions.

It would not be easy to carry out an objective assessment of the differences in appearance of flower pigmentation with respect to temperature or color transformation over time. It would, though, be interesting to discover what mechanism is the primary cause of such differences in appearance and change in color, apart from its being, in the final analysis, of chemical origin.

The hot color range of crocosmias varies from yellow through orange to red, mostly intense, saturated hues. The intensity of the color, or chromatic intensity, is heightened by a glistening or sparkling effect on the surface of the tepals that is more pronounced in hot conditions. The most intense shades of red approach pure red, which means that the flowers can be particularly difficult to photograph, as neither conventional film, for either transparencies or prints, nor digital technology can adequately cope easily with such colors. Attempts to photograph such flowers in close-up can result in a red blur. While infrared automatic focusing of cameras can be ineffective with such hot colors, the blurred results are not purely the result of lack of focus.

Propagation

The corms of most crocosmias increase so readily that simple division and separation of the corms is the easiest method of propagation. While usually only the top corm produces leaves and flower stems, the lower corms are not dead, and if the corms in such a chain are separated and planted, all but the oldest will produce new shoots. In some instances old corms will flower well but produce less foliage, thus giving a better display. For the *Crocosmia* ×*crocosmiiflora* hybrids that produce them, stolons may be detached in spring and potted up. This was the normal method of bulking up new cultivars a century ago, and indeed, freshly rooted stolons were often sold. Detached stolons are grown in warm, close (humid) conditions until they are fully established. Newly established rooted stolons of *C.* ×*crocosmiiflora* cultivars form new corms as they develop and if well grown will flower in their first year. Division of corms is of course the only method for the hybrid cultivars, which do not produce many or any stolons. Most cultivars increase fast enough that separation of corms should be adequate for most nurseries. Micropropagation, though initially unsuccessful, is now being used for some cultivars (Bloom 1991). A few cultivars, such as the ubiquitous *Crocosmia* 'Lucifer' (Plate 13), will come true from seed if self-fertilized, but most will not because of more complex parentage.

All the species can be raised from seed, and many of the hybrids also produce viable seed. In their natural habitat, seeds are subject to being eaten by larvae of bruchid beetles as they ripen inside the capsules. Outside their native habitat, seeds are not attacked by parasites. Fertilization and seed maturation seem to be temperature dependant and are also affected by available moisture when the plants are ripe for pollination. Lower temperatures, probably about 68°F (20°C), are required for fertilization of the earlier-flowering species from higher elevations. *Crocosmia aurea* and the large, late-flowering hybrid cultivars bred from it seem to require a higher temperature, about 77°F (25°C), for fertilization and for a significant part of the gestation period. Nearly every flower of a *Crocosmia* will begin to form seed capsules, whether or not the flowers have been fertilized, but those of unfertilized flowers expand little while fertilized ones enlarge considerably, so there is rarely any doubt as to which capsules will contain seed. False pregnancies do occur, however, where the capsule swells to the size of a full, fertile capsule despite being empty. Relatively high temperatures seem to be required for this phenomenon to occur.

Crocosmias are quite promiscuous, and significant variation in the resultant seedlings may be expected from any cross, even between species. The results of crossing hybrids, especially some of the old cultivars with many generations of inbreeding, may be expected to produce even more variable results. Remaking a known cross between particular cultivars may not produce the same results as the original cross. It is also clear from the descriptions of many cultivars that crossing two large-flowered cultivars can produce smaller-flowered progeny.

Seeds will germinate quite quickly after it is sown if kept moist and at 59–64°F (15–18°C). Seeds may be sown fresh in autumn, as they do not need to be stratified, or the following spring but will lose viability if kept a year or more at room temperature. We do not know whether keeping seeds chilled will extend their viability, as is the case in many plants. Seeds are probably best sown in spring, but if sown in autumn the seedlings require light and heat to be kept in continual growth until late spring. Seedlings of *Crocosmia aurea* and its hybrids with *C. pottsii* can in this way be induced to flower in the first year, if well looked after, as first described by Davison in 1905. The larger species will require 2–3 years' growth before they flower. Where more than one species with concurrent or overlapping flowering periods are grown in close proximity, cross-fertilization may be expected. It is likely that all species will cross-fertilize with one another, though by no means all permutations have yet been tried. There is still scope for creating some distinct and worthwhile new cultivars, not to mention re-creating more large-flowered *C. ×crocosmiiflora* hybrids to replace many of the fine old cultivars now lost to cultivation. A sufficient number of old large-flowered cultivars survive to provide a good basis for such attempts, though as noted, the results may not match expectations. Introducing some of the other species into a breeding program provides the potential for greater diversity. The use of *C. pottsii* seems to be favored commercially because of its prodigious powers of increase, though the use of *C. mathewsiana* might result in better garden plants.

Given suitable climatic conditions, many crocosmias will self-seed, so it is best to remove seed capsules before they are ripe. Unfortunately, *Crocosmia pottsii* is the species most prone to self-seeding. The most popular hybrid cultivar, *Crocosmia* 'Lucifer' (Plate 13), is also known to self-seed, and a significant proportion of plants now sold as 'Lucifer' are seed-raised. Hybrids of this cross, between *C. paniculata* and *C. masoniorum*, seem to be almost as prolific at self-seeding as *C. pottsii*. Curiously, the original Lemoine hybrid, *C. ×crocosmiiflora,* though producing viable seed, may require warmer conditions to reproduce as it does not seem to self-seed readily. Even *C. pearsei* has been known to self-seed, though despite flowering early it does not seem to readily produce viable seed. This may of course be because the agents capable of pollinating the flowers of these crocosmias are lacking.

Perhaps surprisingly, self-seeding of crocosmias, as in many other plants, does not seem to occur within the first few years of being grown in a particular location. A process of acclimatization may be required. Self-seeding of crocosmias seems to normally occur within clumps rather than in the open ground. Such self-seeding is not a serious problem where the foliage of the self-sown seedlings differs from that of the plants among which they have sprouted as they can be easily identified and rogued out even before flowering, if desired. The vigor of *Crocosmia pottsii* can, however, cause problems when it self-seeds within clumps of less robust *C. ×crocosmiiflora* cultivars. If not rogued promptly, *C. pottsii* will overrun and choke more garden-worthy cultivars. Early deadheading of crocosmias before the seeds have fully ripened is the most effective way of preventing unwanted self-sown seedlings.

Pests and Diseases

Both *Chasmanthe* and *Crocosmia* are relatively robust and trouble free in cultivation. They do not appear to be susceptible to any diseases. The only two pests that seem to adversely affect them are red spider mites and thrips.

Red spider mite, the most common problem, particularly for plants grown under glass, also attack plants outside in hot summer conditions. Thrips may also be a problem in milder regions where winters are not cold enough to eradicate this pest. Nothing seems to have changed over the course of the century as these were the only two pests reported in 1907 by Edwin Molyneux. There are pesticides that control both these pests, though changes in regulatory standards may restrict the availability of many well-established remedies. There have been reports of a serious virus infection in the stock of a large commercial nursery in the Netherlands, with large quantities of corms having to be destroyed. Whether the virus originated in or affected only *Crocosmia* is not known. There seems to be a higher incidence of infection of different kinds of plants when the same ground is used for intense cultivation for many generations, or too many plants of the same genus are grown in close proximity.

Mice or voles have been known to eat young corms of *Crocosmia aurea;* they are also very partial to the bulbs of *Crocus,* especially those with yellow flowers. Insect grubs that eat the seeds of *Crocosmia* species inside the developing fruits in their native habitats do not seem to occur outside Africa for capsules of *C. ×crocosmiiflora* in its introduced range around the world are not affected by these parasites. Given such a worldwide distribution of the common montbretia, in the moist temperate and cool tropical parts of the world, they must generally be considered trouble free.

There are no specific known diseases reported for crocosmias. It would be tempting fate to suggest that neither rabbits nor deer find the foliage attractive, but such larger pests do not seem to bother much with crocosmias. In the wild, however, flowering stalks in bloom or with developing capsules are eaten by antelope, suggesting that the plants are not unpalatable. Slugs and snails often seek refuge in the cool moist recesses of the basal leaves of dense clumps of *Crocosmia,* but there is little sign of damage by them.

Cultivars of *Crocosmia*

History of *Crocosmia* Breeding

As outlined in the chapter, Early Exploration and the Discovery, *Crocosmia aurea*, or *Tritonia aurea* as it was initially called, was first introduced into cultivation in England in or perhaps even before 1847. The James Backhouse Nursery in York was credited by Joseph Dalton Hooker as responsible for introducing the plant to British gardens in the text accompanying the 1847 illustration of *T. aurea* in *Curtis's Botanical Magazine*. The species was transferred to the new genus *Crocosmia* in 1851 by Jules Émile Planchon. The form of this variable species that was first cultivated was relatively short, about 2 feet (60 cm) high, with flowers about 2 inches (5 cm) in diameter. A taller and larger-flowered selection was introduced later.

Crocosmia pottsii was evidently first introduced into cultivation in Scotland. One of the hardier species of the genus and well suited to growing in cool moist conditions, it has remained in cultivation ever since. Despite its excessive vigor, it has relatively small red flowers, which do not recommend it for most gardens. Nevertheless, there are at least six distinct clones of the species in cultivation in the British Isles alone. Soon after its initial introduction, *C. pottsii* was obtained by the Victor Lemoine. It was Lemoine who first thought of trying to hybridize this plant with *C. aurea,* and he succeeded in flowering the result of this unlikely union for the first time in 1881. It is uncertain how long *C. aurea* had been in cultivation in France, but Lemoine was clearly very prompt in obtaining *C. pottsii*. The initial cross was made in 1879, using pollen from *C. aurea* to fertilize his new acquisition. As *C. pottsii* flowers much earlier in the season than *C. aurea,* we wonder whether an attempt in 1878 failed and pollen from *C. aurea* was saved from the previous year.

Lemoine's intention in making what he considered to be an inter-generic cross (*Crocosmia pottsii* was then known as *Montbretia pottsii*) was to increase the variety of late summer flowers for display in conservatories or for cut-flower decoration. Neither of the two species would have been hardy outside in winter in northeastern France for the region has a continental climate. Lemoine christened this new hybrid '*Montbretia crocosmiaeflora*', but to conform with the rules of botanical nomenclature the name is now spelled ×*crocosmiiflora* (the multiplication sign × denoting a plant of hybrid origin). Lemoine first listed it for sale 1882. He was apparently more prompt at spreading word of his new hybrid in England than on the Continent. He sent a stem of its first flowering to the editor of *The Garden,* William Robinson, in which the following note was published 21 August 1881 (announcements about the new hybrid also appeared in a second British horticultural publication, *The Gardeners' Chronicle,* in July and August):

> NEW HYBRID MONTBRETIA. Mons. V. Lemoine, of Nancy, sends us a new bulbous plant, which he has obtained by fertilising *Montbretia pottsii* [*Crocosmia pottsii*] (figured in *The Garden* last year, p. 84) with *Crocosmia aurea.* The progeny Mons. Lemoine proposes to name *Montbretia crocosmiaeflora* [*C. ×crocosmiiflora*]. The flowers, he says, are four or five times the size of *M. pottsii;* and this successful cross he considers to be the starting point for the production of a race of beautiful hardy varieties. The flowers sent are borne in the same manner on the spike as those of *M. pottsii,* but they are much larger and of a deeper colour. The cross sent seems to be precisely intermediate between the two parents.

Victor Lemoine was one of the leading European nurserymen of the period, an enterprising businessman and a master hybridizer of plants in many diverse genera. A list of cultivars in numerous genera that he introduced is appended to a brief biographical profile by Masquelier (1995). Lemoine was quick to realize the potential for this vigorous hybrid and immediately began a hybridization program by recrossing the new plant with the pollen of both *Crocosmia aurea* and *C. pottsii,* as was recounted by his son, Émile Lemoine, in 1900. A succession of new cultivars, based on this incestuous breeding program, was released every year after 1882. First came *Montbretia* 'Elegalis' in 1883, then 'Aurea' and 'Pyramidalis' in 1884, and 'Bouquet Parfait, 'Gerbe d'Or' (Plate 15) and 'Étoile de Feu' in 1885. And so on annually (except for 1896) through 1900.

Many of the earliest cultivars were noted as having been discarded by Lemoine in 1900. They were usurped by the subsequent breeding of new, larger-flowered plants. The two large-flowered cultivars of *Crocosmia aurea,* 'Maculata' and 'Imperialis', were brought into the Lemoine breeding program in 1893 (E. Lemoine 1900).

Crocosmia aurea subsp. *aurea* 'Maculata' (Plate 5) was introduced into cultivation in Britain by James O'Brien of Harrow, who was sent corms of the plant from South Africa in 1886. It first flowered in 1888, when it was named *C. aurea* var. *maculata* by J. G. Baker. The later report in *The Garden* noted that, at Kew, the brown spots on the tepals, the feature from which the name was derived, seemed to have lessened in later years (Watson 1893a). The markings do in fact vary, depending on how robustly the plant grows, and are more pronounced in a warm season and when the plant has adequate moisture. When grown exceptionally well, 'Maculata' will produce the brown makings on all six tepals, not just the inner three. To perform at its best, this plant seems to need the optimum combination of heat, moisture and nutrients. O'Brien wrote of this plant in *The Garden* in 1890:

> The history of this beautiful plant is singular. Some years ago a friend in South Africa reported that it had appeared in his garden and whence the corms came from he did not know. Probably it was unwittingly collected with some other bulbs collected wild in some distant part. About 1886 I obtained all the corms he could spare and grew them on. On flowering, its beauty far exceeded my expectations, and I exhibited it at the Royal Horticultural Society (in 1888 I think), and received a First Class certificate. On November 17 1888 it was illustrated in the *Gardener's* [sic] *Chronicle* and soon after Herr Max Leichtlin asked

for and obtained all I could part with. Subsequently, however, I got a few more roots over and again exhibited a small spray this year, sending only a branchlet of the 4-foot [120-cm] spike, as the committee had seen before. The variation in this plant in unaccountable, but it constitutes the first marked departure from the originally figured type, the nearest approach to which is that now called Imperialis, which may, indeed be a slight improvement; but the roots gathered wild and sent to this country many times since the original appeared are very bad forms, although they have to represent the plant in most gardens.

Both *Crocosmia aurea* 'Maculata' and 'Imperialis' were much larger than the original form of *C. aurea* in cultivation. They flowered at almost 4 feet (1.2 m) tall with flowers about 3 inches (7.5 cm) in diameter. Émile Lemoine (1900) credited Max Leichtlin with raising 'Imperialis' by crossing *C. aurea* with itself. He also credited his father, Victor, with naming it and mentions that the plant was self-sterile. William Watson (1893a), assistant curator at Kew, commented on 'Imperialis':

> This *Tritonia* was imported to Kew amongst some plants of *Lissochilus Krebsi,* which were planted in the border in the succulent house, where the tritonia grew and flowered. It was imported by Herr Max Leichtlin about the same time [probably 1891–1892]. Both the varieties as well as the type ripen seeds freely under cultivation.

However, barely a month later, Watson (1893b) corrected this:

> the other is known as Imperialis, and I now learn that this was raised by Herr Max Leichtlin at Baden-Baden, where it was first distributed under the name of Macrantha in 1888. This has flowers even larger than those of Maculata, and like it in substance and color, differing only in the absence of spots. Both varieties grow to a height of three feet [90 cm], and flower freely and continuously all the summer. A third variety, grown at Kew as Imperialis, is evidently distinct from the true plant of that

name, the flowers, although as large, having narrower segments, colored clear yellow. This might be well called Citrina. It is quite as tall at the other two and as free-flowering. There are no other plants of the *Tritonia* section of Iridaceae more useful for the producing brilliant colors out-of-doors in summer than these Crocosmias.

Émile Lemoine was probably correct about his father's proposing a more suitable name for the plant, though it is likely that Max Leichtlin raised *Crocosmia aurea* 'Imperialis' from seed of a larger-flowered genotype (a distinct genetic line) of the species sent from South Africa, where he clearly had good contacts. The introduction of these two new taller, larger-flowered plants into the breeding program provided the impetus for the production of the large-flowered hybrids, as Émile Lemoine (1900) observed. The cultivars released by the Lemoine nursery from 1894 onward had appreciably larger flowers.

The Lemoine nursery was not the only one raising new *Crocosmia* hybrids, as Émile Lemoine graciously noted. Leonard Lille of Lyons, Jacques Welker of Bourgival, Wilhelm Pfitzer of Stuttgart and a nursery neighboring Lemoine, Gerbeaux et Crouse of Nancy, all raised and named a small number of *Crocosmia* cultivars. Little is known of the breeding program of these nurseries, but it is likely to have been based on the early Lemoine cultivars. Several of the cultivars raised by Welker were described and illustrated in the French journal of horticulture, *Revue Horticole* (Grignan 1906, 1907).

A tall but small-flowered species of *Crocosmia* quietly entered cultivation in Britain in 1884, *C. paniculata,* though at this time it was still called *Antholyza* and had been described in 1867. Its existence was, no doubt, overshadowed by the proliferation of the popular new hybrids as there was little mention of it in the horticultural press. Nevertheless, its introduction was reported in *The Garden* in the following short note (Robinson 1884):

Antholyza paniculata—Flowers of this pretty plant have been sent to us by Mr. C. Miles, who collected it in the Transvaal, and in whose garden at Royston it has grown 3 feet [90 cm] high,

bearing *Gladiolus* like leaves, and a long panicle of brown red
and yellow flowers. Mr. Miles says it is growing out of doors and
has increased quite rapidly, and, judging by the specimen before
us, and his description, the plant must be quite at home. The
Antholyzas constitute a small genus of the Iridaceous plants
closely related to the *Gladiolus,* but more like *Tritonia* and
Montbretias than the *Gladiolus.* It is likely to prove a useful
addition to the several good bulbous plants which we already
grow in our outdoor borders, i.e., *Montbretia, Tritonia, Watsonia, Gladiolus* and the old *Antholyza aethiopica* [perhaps *Chasmanthe aethiopica*], which is something like Mr. Miles' plant,
but altogether smaller. Mr. Miles says the leaves of his plant are
serrated. We suspect this means plicated [pleated], as no member of the large order of Iridaceae has serrated leaves.

The van Tubergen nursery seems to have been the first to introduce the
species into commerce when it was listed in their 1895 catalogue. Nevertheless, Max Leichtlin was already growing it in 1892. The subsequent existence of this species in cultivation was barely commented upon. There
are only a scattering of references by one contributor, the keen Scottish
plantsman Sam Arnott, who posthumously had a snowdrop named after
him, so his name is still familiar to many gardeners. The initial brief syn-
opsis of the species that Arnott (1899a, b) provided, and positive comments about growing it in southwestern Scotland, where he gardened, did
not attract any comment by others. Arnott subsequently wrote about a
larger cultivar of *Crocosmia paniculata* called 'Major', which was introduced by Leichtlin in 1900.

In more widespread cultivation, *Crocosmia paniculata* was thought to
be tender despite the introductory note and one in 1899, by Arnott, reporting it as hardy. Periodic catalogue references continued to report it
as being tender. There are, surprisingly, only a very few brief references
to this species in cultivation, and those were mostly made by Arnott
(1900–1910). The form in cultivation, according to Arnott, was from
Natal and had 'deep crimson' flowers, but no full description or illustration of this plant ever seems to have published in the European horticultural press. In 1900 Leichtlin listed a variety *'majus',* without description,

but noting 'it is hardy here', that is, in Baden-Baden. The name implies that this plant was larger than the form he grew in 1892.

Max Leichtlin only produced a plant list and began to sell plants commercially at the end of the 19th century after his finances had seriously declined. Prior to this time, he exchanged plants that he had been the first to introduce into cultivation with other leading gardeners, or given them away. As he clearly supplied a number of nurseries with some of his new rare plants or hybrids from about 1890 onward, however, it is probable that he undertook some occasional commercial dealings before his finances became critical. He not only had a remarkable network of contacts who sent him seeds or plants from all over the world, but he mastered their growing requirements in cultivation before passing them on. From the outset, his horticultural interests were at the forefront of understanding the requirements of new plant introductions and he regularly corresponded with the leading botanists of the period, including Joseph Dalton Hooker at Kew, and Charles Sprague Sargent at the Arnold Arboretum in Massachusetts. He also corresponded and exchanged plants with a veritable who's who of leading gardeners of the period, including Henry Elwes, Ellen Willmott and William Robinson. Many nurseries benefited from his plant introductions and made them available to a wider gardening public. He regularly contributed to *The Garden,* the weekly horticultural journal founded by William Robinson, with his contributions entitled 'Notes from Baden-Baden', a title also used for his contributions to Sargent's equivalent weekly magazine in America, *Garden and Forest.* He typified an age when there were no national or other boundaries between leading plantsmen, be they botanists, gardeners or nurserymen.

Sam Arnott subsequently described Leichtlin's *Crocosmia paniculata* 'Majus' as being taller, later-flowering and with 'scarlet and yellow' flowers. A generation later, a Mr H. Stevens of Chepstow, writing in 1936 in *Gardening Illustrated,* the successor to *The Garden,* described the species as having 'orange red flowers'. This might be the first reference to the hybrid between *C. paniculata* and *C. pottsii,* which is now almost universally grown as the species *C. paniculata* in Great Britain. He went on to recommend that it 'must have sunshine and thoroughly warm and well drained soil', adding that 'it loves a light sandy loam, and generally thrives above chalk'. He also recommended that 'in cold districts annual lifting is desirable, replanting in February'. The following week two of the points made by Stevens, the need for sun and the tender constitution of the plant, were refuted by Sir Herbert Maxwell. He was a leading Scottish gardener whose two books on gardening were an inspiration to the late Graham Stuart Thomas, who in turn became one of the Great Britain's leading plantsmen and authors. Maxwell (1936) noted, 'we plant *Antholyza* [*C. paniculata*], in woodland where it thrives vigorously, and is now [10 October] in full flower, rabbits having no taste for it'.

The burnt orange color form illustrated by Pearse (1978) does not seem to have been in cultivation until much later. Alan Bloom of Bressingham Gardens was growing it in the Dell in the early 1960s (Bloom 1991). The original form described by Arnott may be the red-flowered form also illustrated by Pearse, which has been in cultivation in the British Isles for almost a century. The Daisy Hill Nursery catalogue listed *Crocosmia paniculata,* as an *Antholyza,* in 1906, describing it as having 'deep crimson flower spikes and handsome plaited leaves: a first class hardy plant'. It is likely that it is this red form that was first listed by van Tubergen, but without description, prior to 1895, and that Max Leichtlin was the most likely source. The rare, pure yellow-flowered sport, or mutant, from Mpumalanga province in eastern South Africa, illustrated by Germishuizen and Fabian (1977), is not in cultivation.

Prior to 1891, the van Tubergen nursery listed only *Antholyza aethiopica* 'Major', which is of course *Chasmanthe floribunda*. A third *Antholyza* seems to have been introduced a few years after *A. paniculata*. This was the plant now known as *Crocosmia* ×*crocosmoides,* first listed in 1895 but at several times the price of *C. paniculata*. This would suggest that *C.* ×*crocosmoides* was either slower to increase or was obtained and introduced a year or two later than *C. paniculata*. Neither of these plants increases nearly as quickly as *C.* ×*crocosmiiflora,* so the plants are likely to have been bulked up for a year or two prior to their being listed in the van Tubergen catalogue.

The origin and status of this new *Crocosmia* does not seem to have been widely known. Another new and supposed intergeneric hybrid would normally attract attention and be mentioned in the horticultural press of the time, but this was not the case. Curiously, Sam Arnott does not even mention it, despite his advocacy of the two forms of *C. panicu-*

lata. There is no information in the British horticultural press about where the plant originated or who named it. Almost the first published mention is a rather dismissive comment made by Stevens (1936): 'The older *Antholyza crocosmoides* [*C. ×crocosmoides*] is a fine plant, but scarcely as bright as *A. paniculata* [*C. paniculata*]'. The name *A. crocosmoides,* but without description, is mentioned only twice in the *Journal of the Royal Horticultural Society* in the 100 years following its introduction. Few plants can have been in cultivation for so long and attracted so little attention. In 1898 the Lemoine nursery listed the plant as '*A. crocosmoides* Lemoine', describing it as a 'Beautiful hybrid, with large orange flowers, abundant foliage and 1 meter [3-foot-] tall stems.' The plant, as '*A. crocosmioides*', was included in *The Standard Cyclopedia of Horticulture* (Bailey 1914), but no horticultural encyclopedia in the United Kingdom treated it until later editions of *Sanders' Encyclopaedia of Gardening* (revised by A. J. Macself, about 1930), which clearly relied on Bailey's *Standard Cyclopedia.*

The answer to this anomaly seems to lie in that fact that the only published description of *Crocosmia ×crocosmoides* contemporary with its introduction into horticulture was, surprisingly, in the United States. In 1897, John N. Gerard of Elizabeth, New Jersey, and a regular contributor to *Garden and Forest,* described it:

ANTHOLYZA CROCOSMOIDES.—This plant is said to be a hybrid between *Antholyza paniculata* [*Crocosmia paniculata*] and *Crocosmia* [*C. aurea*]. It is now in flower, and is, I suppose, a production of Herr Leichtlin; at least it was distributed from Baden-Baden garden, where it is said to have proven hardy for several winters. It is an attractive plant, flowering here at a height of about two feet [60 cm]. It has many of the characteristics of the *Antholyza* in its ribbed leaf and many-flowered stems with opposite flowers. These flowers are a bright orange-red, with yellow markings about an inch and a half [4 cm] broad with three inner and three outer segments and a bent tube. While not as richly colored as *C. aurea,* it is an attractive gain if it proves hardy.

The identity of the plant is preserved as three herbarium specimens in the herbarium of the Royal Botanic Gardens, Kew, made from stems supplied by Cornelius Gerrit van Tubergen in late 1904. Two specimens were of flower stems in bud; the third, supplied a few weeks later, had almost finished flowering, with only two open flowers remaining. It is unlikely that Kew would have obtained these specimens if, at the time, the plant was widely known or considered to be a hybrid. It seems surprising that van Tubergen did not know of the hybrid origin of the plant or, if he did know, did not inform Kew. *Crocosmia* ×*crocosmoides* was described again, as a species, *C. latifolia*, by N. E. Brown in 1932. He was apparently unaware that the plant was in cultivation at Kew under the name *Antholyza crocosmoides* and clearly did not know that it already had a validly published botanical name that had appeared in the horticultural literature as a probable hybrid more than 30 years earlier.

Kew had accessioned corms of *Crocosmia* ×*crocosmoides*, as *Antholyza crocosmoides*, from van Tubergen on 17 November 1904 (C. Brough, archivist at Kew Library, pers. comm.). The corms may well have been requested at the time the herbarium specimens were acquired. A large clump of this plant, covering several square yards or meters, still grows at Kew and is almost certainly descended from this accession. It grows in the long, mixed herbaceous bed, backed by a substantial red brick wall, facing the Order Beds. Several other forms of this hybrid survive in cultivation, the best known of which is 'Castle Ward Late', which is similar but about 6 inches (15 cm) taller.

The turn of the century seems to have heralded the rapid decline of commercial interest in hybridizing crocosmias by nurseries in France. E. H. Krelage & Zoon, of Haarlem, the only Dutch nursery that seems to have been active in breeding crocosmias in the 1890s, also seems to have lost interest very early in the new century. Only one other Dutch nurseryman, Jan Roes, seems to have been interested in raising and naming a few cultivars, mostly in the second decade of the 20th century. The Pfitzer nursery of Stuttgart, Germany, sporadically introduced new cultivars from the late 19th to the mid-20th century. It released a fine, large, red-flowered cultivar, 'Germania', in 1899, which proved significant in subsequent breeding in the England.

In the early 1900s the center of breeding shifted from France to Norfolk, England, and interestingly also from commercial nurseries to private gardens. In the British Isles, the role and status of head gardeners in large private gardens had gradually undergone a transformation in the Victorian era. Head gardeners were no longer regarded as just artisans responsible for overseeing the menial garden workforce and keeping the garden in order. Their expertise had become a prized, sought-after commodity. The career of one head gardener, who ended up being knighted, transcended the usually stratified Victorian society—but then, Sir Joseph Paxton was a remarkable man. Head gardeners not only ensured that the ornamental grounds were well looked after but also that there was a regular supply of fresh produce from the kitchen garden for the owners. The ability to grow pineapples was a notable hallmark of a head gardener's expertise. Such produce would also include a regular supply of cut flowers. The various skills of head gardeners in the cultivation of plants, including the forcing of fruit and vegetables, and looking after rare tropical exotics in stove houses or conservatories, were augmented by hybridization skills as well. Many large ornamental and kitchen gardens were not necessarily a self-indulgence on the part of their owners. Often, they were run on a semicommercial basis, with surplus produce and plants sold, to subsidize the cost of running them.

The tradition of competitive showing of horticultural produce, flowers, fruit and vegetables became popular and attracted significant prestige, not only for the garden owners but also for the head gardeners. Novelty was then, as it always seems to have been and still is, a much sought after commodity among the gardening public. Gardening was no longer the preserve of the aristocracy but expanded to include wealthy businessmen and the more or less affluent middle classes. The regular competitive shows mounted by the Royal Horticultural Society in London were the most prestigious in Great Britain. The judging committees could give worthwhile plants an Award of Merit or the even more prestigious First Class Certificate. Plants usually had to be shown at least twice before being given an award, with only a Certificate of Preliminary Commendation given to exceptional plants on their first showing. Plants that particularly impressed the judges could and often were given Awards of Merit on their first showing. Very few exceptional plants received a First Class Certificate on their first outing, but this breached the normal convention or rule that plants first had to have an Award of Merit before they could be considered for a First Class Certificate. The various Royal Horticultural Society floral committees seemed to exercise some latitude in their duties occasionally. (In more recent years the Royal Horticultural Society introduced a different award called the Award of Garden Merit. This award signifies that the plants are deemed to be garden-worthy, that is, robust, attractive and not requiring special care to perform well in the garden. Not having an Award of Garden Merit is not necessarily a negative reflection on the plants; it indicates, most likely, that there has been not yet been an opportunity to assess them.)

George Davison was head gardener at Westwick Hall in Norfolk. He was the third generation in his family to hold the post, with a son being groomed to follow him ('a Yorkshireman' 1906). Davison (1905) published an article, 'The new montbretias', describing their attractiveness and decorative use as well as their cultivation requirements. He mentioned some of the new cultivars, discretely giving fuller descriptions of those that he had raised and named but without mentioning his hybridization program. It was left to Molyneux (1907) to describe how he bred his greatly improved hybrids. Davison's plants were released commercially through the well-known nursery Wallace and Company, no doubt through some early form of marketing agreement. The garden at Westwick Hall also included a large orchard, which seems to have been run commercially.

Fortunately, Edwin Molyneux, awarded the Victoria Medal of Honour by the Royal Horticultural Society, saw fit to record and publish an account of how the first major breakthrough in the development of crocosmia breeding came about. The common montbretia, *Crocosmia* ×*crocosmiiflora*, already grew at Westwick Hall. George Davison obtained 'Golden Sheaf', one of the earliest Lemoine cultivars, which he crossed with *C.* ×*crocosmiiflora* in the hope of obtaining a larger-flowered yellow plant. His initial results were disappointing. Using one of the progeny with large but dull-colored flowers, he recrossed it with 'Golden Sheaf'. The result was satisfactory and flowered earlier than the parents, in July rather than August. Named after himself, 'George Davison' was exhibited at the Royal Horticultural Society in 1902 and obtained an Award of Merit. In 1901 he acquired 'Germania', the new Pfitzer cultivar—the result of pollinating a Lemoine cultivar, 'Étoile de Feu', with the pollen of *C. aurea* 'Imperialis'; 'Germania' had very large, bright red flowers on well-branched 4-foot (120-cm) stems—which was given an award of merit (*Wertzeugnis*) in 1900 at Frankfurt, Germany, and an Award of Merit from the Royal Horticultural Society in 1901.

Davison used the pollen of 'Germania' in his breeding program and named two of the resulting seedlings 'Hereward' and 'Ernest Davison', which had large flowers in varying shades of orange and bloomed late in the season. Several other cultivars followed before one, 'Westwick', was crossed with 'George Davison' to produce 'Prometheus' (Molyneux 1907).

The large, rich golden-flowered 'Prometheus' (Plate 15) had a maroon eye surrounded by a broad maroon halo. It was widely hailed in the gardening press of the day as a significant breakthrough in *Crocosmia* breeding and was illustrated in many publications with varying degrees of accuracy. The best one seems to have been in the *Proceedings of the Royal Horticultural Society* 30: 141, figure 44, in 1906, which recorded that the plant received an Award of Merit. Most *Crocosmia* cultivars were never illustrated; those few that were normally only featured in a single publication and then often in a group of cultivars. Prior to this, the only cultivar that had been illustrated individually was Lemoine's original hybrid, *C. ×crocosmiiflora.*

Several other cultivars followed, including 'Comet', before Davison reached the culmination of his breeding program with 'Star of the East' (Plate 13), raised in 1910. The quality, substance and habit of this cultivar, which has flowers as much as 4 inches (10 cm) in diameter, has never been surpassed. It received the very unusual accolade of being awarded a First Class Certificate from the Royal Horticultural Society in 1912, when it was first shown.

Perhaps Davison realized that he was unlikely to improve on this cultivar, as he subsequently ceased hybridizing crocosmias and, instead, turned to breeding apples. The cultivation of apples and pears was a noted feature of the garden at Westwick Hall. They were grown in quantity for sale, so economics may have played a part in the significant change in gardening emphasis. His employer, Major Petre, was still a commissioned officer, subsequently rising to the rank of colonel, so this may also have influenced circumstances. When Davison gave up hybridizing crocosmias, stock of his breeding program was passed to another garden, that at Wreatham Hall in Norfolk.

Wreatham Hall was owned by Sidney Morris; his head gardener was George Henley. The house itself had been destroyed by fire many years earlier, but the garden was still being maintained. Henley had been active in hybridizing crocosmias for some time, and there was an overlap of several years when both Davison and Henley were breeding new crocosmias. Henley achieved success early on when a crocosmia cultivar, which he named after himself, was given an Award of Merit in 1909 at the same time as 'Pageant', bred by Davison, was similarly honored.

In about 1912 or 1913, Morris moved to another property in Norfolk, Earlham Hall, where he commissioned the firm of Wallace and Company to design and lay out a new garden (Bowles 1917). On completion of the work, one of their employees, J. E. Fitt, who had supervised the work, was asked to stay on as an assistant to Henley, who was nearing retirement. John E. Fitt, known as Jack, took over as head gardener at Earlham Hall several years later, on the retirement of Henley in about 1918, just before the end of the First World War.

New cultivars were regularly exhibited at the Royal Horticultural Society each year, with a succession of plants being given Awards of Merit: *Crocosmia* 'Queen Adelaide' in 1913, 'Queen Elizabeth' in 1915 and 'Queen of Spain' in 1916. After skipping a year, Awards of Merit were given to three new plants in 1918—'Queen Alexandra' (Plate 14), 'Queen Mary' and 'Nimbus' (Plate 14)—and the following year 'Una' was given an Award of Merit and 'His Majesty' (Plate 14) gained a First Class Certificate on its first showing, following the precedent set by George Davison's 'Star of the East' (Plate 13). These were likely to have been the last cultivars in which George Henley played an active part. Two more awards followed the next year for 'Joan of Arc' and 'James Coey', named after the proprietor of the Slieve Donard Nursery in Northern Ireland, 2 years before he died. This brought the total of Wreatham Hall–Earlham Hall hybrid awards to 10 in 1920, an average of one per year since 'George Henley' set the precedent in 1909.

The program of *Crocosmia* hybridization was apparently not affected by either the move to Earlham Hall or the war, and the succession of a new head gardener ensured that it continued with undiminished momentum. The tour de force of 1918–1919 was followed by an annual output of good new cultivars for well over a decade, with regular showing at the Royal Horticultural Society and equally regular awards. The plants raised by Jack Fitt and George Henley subsequently became known collectively as the Earlham Hybrids according to Phillip Wood (pers. comm. 1994) of Slieve Donard Nursery, now retired. The term seems to have been used for all the large-flowered hybrids, including those raised prior and subsequent to Jack Fitt's tenure at Earlham Hall, and probably by default to all large-flowered *Crocosmia* cultivars, irrespective of where they were raised or by whom.

The prestige attached to plants given awards by the Royal Horticultural Society not only bolstered the esteem of the gardener and garden owner but also accentuated the demand for such plants. In line with market forces, this ensured that a high price could be obtained for the initial release of such plants. The new plants were sold directly from Earlham Hall, and several such plant lists survive. The first release of corms of *Crocosmia* 'His Majesty' (Plate 14) in the autumn of 1920 attracted a premium price of 2 guineas (£2 2 shillings or £2.10, or $3.50, the equivalent of well over £20 or $30 today) for a dry corm, rather more than the average weekly wages for a gardener of the day. Plants 'in spring', a euphemism for rooted stolons, were priced at 15 shillings each (75p, or $1.25), about a third the price of a corm. Other cultivars, available from the same list, were significantly cheaper but still relatively expensive. They, too, were sold as 'plants in spring' though many were also available as corms at 1.5–2.5 times the price. A rooted stolon would produce one flowering stem the same year; a corm could produce probably three flowering stems or possibly more in the case of 'His Majesty', which is quite vigorous despite having small corms.

On the death of Sidney Morris in 1924, Jack Fitt inherited the entire *Crocosmia* stock. He subsequently moved to Breccles Hall in Norfolk, where he was appointed head gardener to the Hon. Mrs Edwin Montague. *Crocosmia* breeding there continued unabated until the early 1930s (Pea 1931). The names of many of these later cultivars reflect the horticultural and social circles in which Beatrice Venetia Montague moved. Crocosmias continued to be sold directly from the garden, this time by the dozen and still under the title Earlham Giant Montbretias. The prices were expensive and still well beyond the reach of many private gardeners. They ranged from 10 to 36 shillings (50p to £1.80, or $0.85–3.00) per dozen. Commercial nurseries probably constituted the majority of customers as they would have regarded the corms a worthwhile investment. Though not included in the two known surviving price lists, corms of unnamed seedlings were also sold to nurseries, which sold them as 'mixed seedlings' for rather less than the named cultivars.

Unfortunately, there is no complete record of the Earlham Hybrids, so it is impossible to know just how many cultivars were named. Two breeding notebooks of J. E. Fitt survive and are now in the care of Norwich City Council; they were presented by Peter Fitt, a son of Jack Fitt. These notebooks contain descriptions of some unnamed seedlings and several named ones, which never seem to have been released commercially, or if they were the names were changed. It is likely, for instance, that the name 'Red Ensign' was changed to 'James Coey' when it was released commercially. A similar fate seems to have befallen 'Macaw'. The editor of *The Garden* noted that 'Halo', 'Macaw' and 'Red Indian' were named during his visit in September 1917; however, only the first and last names occur later, as cultivars raised by George Henley (Cowan 1917). Subsequently, the converse is also true, in that several nurseries listed cultivars attributed to J. E. Fitt, but they are not recorded in the surviving notebooks. Fitt was clearly still raising, naming and exhibiting some *Crocosmia* cultivars well into the 1930s, but the demand was past its peak, and many bulb catalogues of the period show a marked reduction in the number of crocosmias offered, though several new cultivars were listed.

The demise of this great age of *Crocosmia* breeding was hastened by the lean and bitter years heralded by the Wall Street crash and concluded by the reconstruction following the Second World War. Changed circumstances altered priorities and led to a fairly rapid decline in labor-intensive gardening. A belated, brief article on montbretias, their cultivation and some of the best sorts, by Fitt (1932) may have been an attempt to counter the prevailing conditions. A number of nurseries still listed a good but reduced range of crocosmias throughout the 1930s and even the 1940s. There was only an occasional mention of *Crocosmia* in the horticultural press from the late 1920s through the late 1930s. These brief comments tended to emphasize the tender nature of the recent large-flowered cultivars, without mentioning the Earlham Hybrids by name. These fairly repetitive articles tended to be primarily recommendations to lift the corms in autumn and to give advice on winter storage and replanting in February. The Second World War did not bring amenity horticulture to an absolute halt, when through necessity there was a great increase in the growing of vegetables and fruit. Ornamental plant nurseries still produced substantially reduced catalogues during the 1940s, and a few crocosmias were included in them. There was only one brief mention of them in *The Gardeners' Chronicle* in 1947, after a 10-year lapse, and nothing further for more than a decade afterward.

The early 1950s saw the release of two new cultivars, *Crocosmia* 'Emily McKenzie' (Plate 15) in England, which received an Award of Merit from the Royal Horticultural Society in 1952, and 'Carmin Brilliant', released the Pfitzer nursery in Germany in 1953, though this might have been 'Carminea' renamed. Despite postwar austerity in Europe, amenity gardening was slowly reestablished. The pleasure of growing brightly colored flowers was, no doubt, considered to be an antidote to the prevailing circumstances when most commodities were in short supply.

It was another decade before any further serious attempt to hybridize *Crocosmia* species occurred. Following the severe European winter of 1963, with deeply penetrating frosts and no snow cover, Alan Bloom discovered that both *C. masoniorum* (given an Award of Garden Merit by the Royal Horticultural Society) and *C. paniculata* proved to be hardy. This provided the impetus for him to try hybridizing these and other species of the genus, though at that time *C. paniculata* was still regarded as belonging to the genus *Curtonus,* so he thought that he was crossing plants from closely related genera. Three species were involved, *Crocosmia masoniorum, C. paniculata* and *C. pottsii,* and the old hybrid montbretia, *C. ×crocosmiiflora.*

As Bloom (1991) relates, the hybridization was rather random, or 'fiddled' as Percy Piper, one of his most experienced foremen, used to term it. Several hundred seedlings were raised and 2 years later the initial process of selection was carried out, reducing the number of plants to 25, with only 6 of these finally selected the following year. Subsequently, another 2, from the original group of seedlings, were named.

The most famous of the Bressingham crocosmias is, without doubt,

'Lucifer' (Plate 13), which is by now almost equal in its worldwide popularity and distribution to Lemoine's original hybrid, *C. ×crocosmiiflora*. In large measure, this one cultivar is responsible for generating renewed interest in *Crocosmia*. Curiously, Bloom seems not to have followed up on the initial success of this group of cultivars. He also seems to have been unaware that the many of the old large-flowered *C. ×crocosmiiflora* cultivars were still in existence, or perhaps was less interested in them because of their reputation for not being hardy. Many of the old cultivars were apparently still readily available in the 1950s and 1960s, and only seem to have been lost to cultivation quite recently. In the early 1960s, the van Tubergen nursery still listed a good selection, as did both the Daisy Hill and Slieve Donard Nurseries in Northern Ireland. Thus it is probable that some of the old cultivars still survive in old gardens in milder parts of the British Isles and have yet to be rediscovered. One of us (G.D.) has managed to rediscover a few and to identify many others that had subsequently been given alternative names.

In the 1980s two other breeders were at work. In England, Phillipa Browne, while working for the nursery Treasures of Tenbury, raised a number of cultivars, mostly as the result of crossing *Crocosmia* 'Jackanapes' with 'Solfatare' (both, Plate 14). She named a total of about 10 cultivars, the best known of which is 'Dusky Maiden', with bronze foliage and burnt orange flowers, which is a distinct break from the more usual, vibrantly colored crocosmias.

It is surprising, given the scale of the Dutch bulb trade, that apart from a few cultivars raised and named early in the 20th century there was so little activity in the hybridization of crocosmias in the Netherlands until relatively recently. Apart from the Krelage nursery, already mentioned, the only other nursery in the Netherlands that seems to have been active in hybridizing crocosmias early in the 20th century was that of Jan Roes, who raised a few cultivars mostly between 1910 and 1920. It would seem that there was virtually no breeding activity with crocosmias in the Netherlands for more than half a century. Other genera may have taken priority, notably *Crocus* and *Tulipa*.

The nursery of Jac M. van Dijk raised quite a few cultivars in the 1980s and 1990s, mostly small-flowered and vigorous forms of *Crocosmia ×crocosmiiflora*. The parentage of some of them was purported to be a few of the old large-flowered cultivars, but this provenance seems unlikely given the poor caliber of some of the plants raised and named. Some would probably be difficult to distinguish from the variety of plants regarded as common montbretia. Many nurseries in the Britain still list 'James Coey' though it is almost certainly no longer in cultivation and is most likely 'Carmin Brilliant', which is currently available commercially. Many of the more recent Dutch cultivars were released through the Dutch wholesale bulb trade, resulting in inconsistent and unreliable naming. Some of the corms supplied under cultivar names are merely poor forms of the common montbretia, and indeed, Dutch catalogues tend to separate out the better plants, listing them as *Crocosmia,* while the poorer named forms are listed as *'Montbretia'.* It is likely that many of the early Lemoine cultivars would probably today also be regarded as being almost indistinguishable from the common montbretia. As there is quite a variation in the surviving montbretias, be they in old gardens or garden escapes, it is likely that some of these forms are Lemoine cultivars. Several of his early cultivars do still survive, and it is not beyond the bounds of possibility that the Pfitzer cultivar 'Carmin Brilliant' (or 'Carminea') was also one of them.

Descriptions of many of the van Dijk cultivars are published by the Koninklijke Algemeene Vereeniging voor Bloembollencultuur (KAVB, Royal General Bulb Growers' Association; van Scheepen 1991). The descriptions are, however, not specific enough to be definitive. In the 1990s the van Dijk nursery closed and its stock was bought by another Dutch nursery, Kwekerij Davelaar. Some of the stock was found to be virus-infected and had to be destroyed. The remaining plants were partially mixed and unidentified, so many of the surviving plants were renamed. A more rigorous attempt to stabilize the naming of specific cultivars is being made by the current owner. Several other nurseries in the Netherlands are active in growing crocosmias. At least one nursery is raising new hybrids from seed based on crossing *Crocosmia masoniorum,* but these seem either to be sold under wrong cultivar names or simply as the species, adding further confusion to the naming of cultivars. At least one worthwhile such plant, bought in quantity by Bridgemere Nurseries, Cheshire, England, was found to be not true to name; it was thus named *Crocosmia* 'Highlight' by Chris Saunders (pers. comm.) of Bridgemere Garden Centre in 1998 before being put on sale.

Since the late 1980s, several others have been active in breeding crocosmias in the British Isles, often naming only a few cultivars. The late John A. Hogan from Cornwall raised two good red cultivars. Terry Jones from Zeal Monochorum in Devon raised three. Alan Lewis, when based at Ford Abbey in Somerset, raised and named five. The only remaining National Collection holder, David Fenwick, has bred and named at least one crocosmia and has also named several other cultivars that he obtained from various individuals. Jim Mahir in Dublin, Ireland, has raised and named about nine cultivars. One of us (G.D.) has raised and named in excess of 10 cultivars, involving almost all the species in cultivation. In most instances, seeds from open-pollinated plants were used, but several specific crosses were made, most notably using the pollen of *Crocosmia pearsei* to fertilize *C. masoniorum* 'Rowallane Yellow', which produced a surprising variation in the resulting plants, in both color and form, but including one with intermediate floral characteristics. The reverse cross was also made at the same time, but the resulting seedlings took a further year to flower and were consistently shorter; the influence of *C. masoniorum* is dominant in all the resulting plants.

Crocosmias have gained in popularity since the early 1990s, so it is likely that more breeding will take place. There is still ample scope to create distinctive and worthwhile new cultivars. Improved hardiness, subtler coloring, longer-lasting flowers and reduction of excessive vigor of vegetative increase are challenges still to be met. The remarkable variety that was obtained from two species in more than 50 years of inbreeding can only hint as the possibilities with further interspecific hybridization. Two more recently raised cultivars provide an insight into future developments. One, 'Fugue', a complex hybrid involving *C. paniculata* and *C. masoniorum,* or a hybrid with dominant features of this latter species, is a tall sturdy plant with five to six compact panicles of large flowers arranged vertically and opening almost simultaneously, giving the impression of a very substantial inflorescence. The other, 'Quantreau' (*C. ×crocosmiiflora* cultivar × *C. masoniorum*), has a stout stem bearing large orange flowers of great substance, apparently typical of *C. ×crocosmiiflora* but with relatively short pleated foliage.

The story of the *Crocosmia* cultivars does not end with those who raised them. Most of the cultivars are now lost to cultivation. Given the relatively tender nature of the corms, it is perhaps surprising that so many of the old cultivars have survived. The persistence of many crocosmias may in part be the result of chance and favorable circumstances but it is also due to a number of keen gardeners in succeeding generations who collect, grow and pass on worthwhile plants. Unfortunately, the identity of many of these gardeners is not known.

Gardening tends to move in cycles with particular plants moving into and out of fashion. Economics ensures that nurseries follow such fashions and stop propagating and stocking plants for which there is little or no demand. A significant proportion of old cultivars of many genera do not survive such fashion cycles; those that do are scattered in old gardens, perhaps unidentified and often all but unnoticed. Margery Fish did much to regenerate interest in old cottage gardens, where many worthwhile and usually robust plants survived. Her interest apparently did not extend to crocosmias as she only mentions growing 'Solfatare' (Plate 14), albeit under her unique spelling, 'Solfaterra' (Fish 1965, 1966).

Graham Stuart Thomas, one of England's greatest plantsmen and who died in 2003, seems to have been the first in recent times to take an interest in this neglected genus. As Gardens Advisor to the National Trust, he was well placed to see the vast range of plants that grew, and in some cases survived, in the diverse range of large gardens that gradually came into the care of the trust.

It is perhaps ironic to note that the majority of visitors to National Trust properties now go to see the gardens rather than the buildings. The trust was originally set up to save and preserve houses of architectural merit and historic interest in the United Kingdom. Many had large gardens, which had to be taken on by default and cared for, as they could not be separated from the houses, for which they provided the appropriate setting. It was only through the indefatigable efforts of Vita Sackville-West of Sissinghurst that the garden of Hidcote became the first garden to be taken into the care of the trust. *Crocosmia* 'His Majesty' (Plate 14) somehow survived at Hidcote, though its planting in that garden was probably due to the efforts of the elusive Nancy Lindsey rather than Major Lawrence Johnston.

Graham Stuart Thomas rediscovered and identified quite a few old cultivars in various National Trust gardens around the country and established a collection of them in the garden at Lanhydrock in Cornwall. He added to it other good cultivars, such as *Crocosmia* 'Nimbus' (Plate 14) and 'Mrs Geoffrey Howard' (Plate 15; though under the name 'Vesuvius'), which the late Norman Haddon had grown in his garden at Porlock. This became the first significant collection of crocosmias, under the care of the head gardener there, Peter Borlase, and subsequently Tommy Teagle. It became the first National Collection of crocosmias, under the auspices of the National Council of the Conservation of Plants and Gardens, which was established in 1978 to promote the collection and conservation of plants, primarily garden cultivars, so that they would not be lost.

Other good old and new cultivars came to light with a trial of crocosmias by the Royal Horticultural Society in its garden at Wisley in the early 1980s. Several more old cultivars were supplied to the trial by the late David Shackleton. He had built up one of the largest collections of herbaceous plants in the British Isles in his garden, Beech Park, southwest of Dublin. There, among others he grew 'Croesus' (Plate 13), 'Hades', 'His Majesty' (Plate 14) and 'Mrs Geoffrey Howard' (Plate 15). Old gardens in Ireland seem to have been a safe haven for many old *Crocosmia* cultivars, though where some of them were rediscovered is not known. *Crocosmia aurea* subsp. *aurea* 'Maculata' (Plate 5) survived in the inner walled garden of Malahide Castle, County Dublin, after the death of Tylo de Malahide, long after many other choice plants had gone. *Crocosmia* 'Croesus', which Shackleton grew under the provisional name of 'Mr Bedford's', came from the garden at Straffen, County Kildare, which is famous for the snowdrop named after it. David Shackleton's son Jonathon, and his wife, Daphne, continue the family tradition in their new garden, Lakeview House, County Cavan. Two other famous Irish gardeners, Helen Dillon and Jim Reynolds, grow and distribute some of the good cultivars from their well-known and much-visited gardens.

Two old cultivars survived in the garden of Castle Ward House, County Down. This property, now in the hands of the National Trust, was the country seat of Lord and Lady Bangor. The late Lady Bangor was apparently attracted by crocosmias at an early age. She brought back corms of a small-flowered yellow cultivar, probably 'Sulphurea', from France in 1900, when she was only 18 years old (William Lennon, pers. comm.

about 1985). It is the tallest and most robust of about six different clones of this small-flowered cultivar. The estate gave its name to a rather different hybrid, between *Crocosmia aurea* and *C. paniculata,* raised by Max Leichtlin, that is now known as *C. ×crocosmoides* 'Castle Ward Late'. This plant derives its present name from both its late-season flowering and the site of its rediscovery. It was the late Molly Sanderson, best known in alpine gardening circles on both sides of the Atlantic, who reintroduced this hybrid into cultivation under the name 'Castle Ward Late' and distributed it with her legendary generosity. It is the one of five different clones of *C. ×crocosmoides;* another is 'Vulcan' (Plate 13), the tallest of all, with flowers that are distinct from the others, being radially symmetric. That cultivar was rediscovered in an old garden in Belfast by the late Ronnie Cameron (pers. comm. 1994) and identified from a 1906 Daisy Hill Nursery catalogue description.

A number of old cultivars also survived at Mount Stewart garden, with its favorable, almost subtropical microclimate, in County Down. Several of these were rediscovered there by Michael Wickeden, now of Cally Gardens Nursery, Scotland, when he worked there as a gardener. He was the first nurseryman in recent times to put some of the old *Crocosmia* cultivars into commerce.

One of the us (G.D.) began growing and collecting crocosmias in the late 1980s and gradually built up the largest cultivated collection of both species and cultivars, with the assistance of those just mentioned and many others. Many more old cultivars were rediscovered, and most have been identified, including 'E. A. Bowles', 'Croesus' (Plate 13), 'George Davison', 'Hereward', 'Jessie', 'Lady Oxford', 'Norvic', 'Mephistopheles' (Plate 15), 'Prometheus' (Plate 15), 'Queen Alexandra' (Plate 14), 'Queen Charlotte', 'R. W. Wallace', 'Sir Matthew Wilson' (Plate 13), *Crocosmia pottsii* 'Grandiflora' and 'Vulcan' (Plate 13), derived from *C. ×crocosmoides.* The collection of crocosmias, along with two other genera, were given National Collection status in the early 1990s. A substantial amount of information about *Crocosmia* cultivars and their breeding was accumulated, initially with the assistance of Dr Margaret Andrews and subsequently Jill Hutchens, both of the National Council of the Conservation of Plants and Gardens, as well as many others, and also through numerous visits to the RHS Lindley Library.

A couple of years subsequent to becoming a National Collection holder, G.D. was contacted by David Fenwick. Fenwick had begun to collect and name 'garden escape' montbretias from the outskirts of Plymouth, England, and had become very interested in *Crocosmia*. Over a couple of years G.D. supplied him with more than 80 different crocosmias, including most of the better old cultivars that were not readily available and copies of the substantial amount of information about the breeding history and cultivar descriptions. David reciprocated by exchanging the forms of montbretias that he had collected and additional information that he had obtained. Once his collection was large enough, Fenwick's application for National Collection status was supported by G.D. It was initially turned down as it had not been established for a sufficiently long period but was subsequently granted probationary status for a year, after which full National Collection status was granted.

From the outset, Fenwick adopted a high-profile approach in the promotion of his collection in an attempt to find further other old cultivars but with little success to date. He did, however, make contact with Peter Fitt, a son of Jack Fitt, who had saved extensive information about his father's breeding work, including two breeding notebooks that were presented to Norwich City Council. David has done much to promote the genus with his Web site, which provides information and illustrations of the flowers. His indefatigable enthusiasm has paid dividends in Norfolk, where Norwich City Council has established a *Crocosmia* garden. Its purpose is to celebrate the dominance of the county in the breeding of crocosmias in both the early and late decades of the 20th century. Both Dunlop and Lanhydrock have withdrawn from the National Collection scheme, so Fenwick currently holds the sole National Collection.

Descriptive List of Cultivars

Crocosmia cultivars are presented in alphabetical order (except that the hybrids *C. ×crocosmiiflora* and *C. ×crocosmoides* are listed first), and the list includes date, name of the breeder and parentage when known. Unless the parentage is stated, all cultivars are forms of *C. ×crocosmiiflora*, that is, *C. pottsii* × *C. aurea*. Apart from *C. ×crocosmoides*, that is, *C. paniculata* × *C. aurea*, no other species was involved in any hybridization program prior to the raising of the Bressingham cultivars in the late 1960s. The parentages of a number of Earlham Hybrids, raised by George Henley and Jack Fitt, represent information from the notebooks of J. E. Fitt, courtesy of David Fenwick's Web site, 'Norwich in Bloom'.

In general, the date refers to the likely date of naming; it has been assumed that when plants were first listed or shown for award, they were named the year before they were listed or shown, unless there is information to the contrary. It is unlikely that a new cultivar could be presented at its best the first year it flowered, though there are documented exceptions. In a number of instances, the same cultivar name was used by more that one breeder, sometimes contemporaneously. Where a cultivar is of likely, but uncertain, attribution to a particular breeder, this is indicated by a question mark following the breeder's name.

Original catalogue descriptions are given primacy in the descriptions but are not normally referenced except where relevant, that is, when the catalogues are those of the raiser of the plant. Where original descriptions are very brief and more extended descriptions were published later, usually by others, the subsequent published descriptions are included and referenced.

Many of the descriptions should not be taken too literally, particularly as regards size of flower and height. Before 1900, 2 inches (5 cm) was regarded as large for a *Crocosmia* flower. Some descriptions are flattering or exaggerated, and the comparisons with other cultivars and dimensions were sometimes clearly subjective. Optimal cultivation and extremely rich feeding doubtless created plants somewhat larger than those grown in average conditions. There is much variation in the descriptions of the flowers, their color and size, and the height to which plants grew when a number of sources are compared. The color terminology encountered is subjective, vague and thus variable.

Awards given to plants by the Royal Horticultural Society (RHS) and Koninklijke Algemeene Vereeniging voor Bloembollencultuur (KAVB, Royal General Bulb Growers' Association, the Netherlands), with dates, are listed at the end of the description. Award descriptions, where adequate, are used in preference to other ones as they are likely to be the most objective.

In roughly chronological order of their activity (dates approximate) with *Crocosmia,* the breeders or introducers mentioned in the list are Victor Lemoine, nurseryman in Nancy, France, 1880–1905; Max Leichtlin, plantsman in Baden-Baden, Germany, 1888–1891; François-Valerie Gerbeaux, nurseryman in Nancy, France, 1884–1910; James O'Brien, Harrow, England, 1880s; Leonard Lille, nurseryman in Lyons, France, 1890s; Jacques Welker, nurseryman in Bourgival, France, 1890–1906; Wilhelm Pfitzer nursery, Stuttgart, Germany, 1890–1954; E. H. Krelage & Zoon, nurserymen in Haarlem, the Netherlands, 1892–1904; Gorge D. Davison, Westwick Hall, Norfolk, England, 1898–1912; Mooy Polman, nurseryman in Haarlem, the Netherlands, 1904–1910; Roes & Goemans, nurserymen in Vogelenzang, then Bennebroek, the Netherlands, 1905–1913; P. S. Hayward, nurseryman in Essex, England, 1912–1913; Jan Roes, nurseryman in Heemstede, the Netherlands, 1913–1925; Thomas Smith, nurseryman in County Down, Northern Ireland, 1914–1915; George Henley, head gardener, Earlham Hall, Norfolk, England, 1907–1918; John E. Fitt, head gardener, Earlham Hall, Norfolk, England, 1919–1940s; C. R. A. Hammond, gardener in Norfolk, England, 1930s; William Slinger, proprietor of the Slieve Donard Nursery, Northern Ireland, following the death, in 1921, of James Coey, 1930s; Albert E. Hill, gardener in Derbyshire, England, about 1932; K. McKenzie, gardener in Northumberland, England, 1950s; J. van Nieuwkoop, nurseryman in the Netherlands, 1950s; Alan Bloom, nurseryman in Norfolk, England, 1965–1980; David R. Tristram, nurseryman in Sussex England, 1965–2002; Eric Smith and Jim Archibald, nurserymen in Dorset, England, 1970–1975; Mrs C. J. M. Ursem, private gardener (?) in the Netherlands, 1980s; J. Zonneveld, nurseryman (?) in the Netherlands, 1980s; Jac M. van Dijk, nurseryman in Haarlem, the Netherlands, 1980–1990s; Phillipa Browne, nurserywoman in Worcestershire, England, 1980–1990s; Percy Picton, nurseryman in England, 1985; Vere Cattermole and Paul Durrant, nurserymen in Wales, late 1980s; Beth Chatto, plantswoman and nurserywoman in Essex, England, 1990s; John A. Hogan, private gardener in Cornwall, England, 1990s; Terry Jones, plantsman at Zeal Monochorum, Devon, England, 1990s; Alan Lewis, nurseryman based at Ford Abbey, Somerset, England, 1990–2001; Christopher Lloyd, plantsman and gardener in Kent, England, 1990s; Jim Mahir, part-time nurseryman in Dublin, Ireland, 1990s;

Kenneth Regney, South African now residing in Hampshire, England, 1990; Chris Saunders, nurseryman in Cheshire, England, 1990s; P. Knuckley, nurseryman in Cornwall, England, 1992; Gary Dunlop, gardener in County Down, Northern Ireland, 1993–2002; David Fenwick, U.K. National Collection Holder of *Crocosmia*, Devon, England, 1994–2002; Sheila Harding, nurserywoman in County Meath, Ireland, 1995–2000; Willem Heemskerk, nurseryman in Davelaar, the Netherlands, late 1990s; Dan Hinkley, plantsman and nurseryman in Washington state, United States, late 1990s; and Ray Hubbard, nurseryman in Devon, England, late 1990s. Cultivars that cannot be attributed to a particular breeder are also indicated. Synonymous cultivar names are also discussed and cross-referenced to the correct name when possible. We point out that many cultivars are no longer extant, possibly as many as half of those listed here. Many others are not readily available in the horticultural trade. The availability of cultivars changes from year to year and with different suppliers.

> *Crocosmia* ×*crocosmiiflora* (Plate 12), 1881, Lemoine, *C. pottsii* × *C. aurea.* This seminal hybrid is intermediate between the parents. It has broad, trumpet-shaped orange flowers with tepals that flare widely open; there is a ring of small V-shaped crimson markings around the center. RHS First Class Certificate, 1883.
>
> *Crocosmia* ×*crocosmoides,* about 1890, Leichtlin. *C. paniculata* × *C. aurea.* Fine, yellow flushed red, starry flowers on branching panicles held above narrow pleated foliage; late flowering (Gerard 1897).
>
> 'Achilles', 1919, Fitt. 'Lord Nelson' × 'Pageant' (parentage from J. E. Fitt notebooks). Tallest of the montbretias, growing to more than 53 inches (1.4 m). More than 130 flowers and buds of a fine rich dark color have been counted on one branching stem. New but not yet exhibited according to Fitt (1920).
>
> 'Adonis', 1919, Jan Roes. Listed in 1923. Clear yellow inside, outside red.
>
> 'A. E. Amos', 1926, Fitt. Beautiful, growing about 42 inches (1.1 m). Deep fiery orange flowers shaded gold. Good branching habit (Fitt 1928).

'**African Gold**', 2000, breeder unknown. Medium-sized, golden yellow-flowered, very similar in appearance to 'Buttercup'. It is likely to be of Dutch origin.

'**Aisha**', 1997, Mahir. Plants about 18 inches (45 cm) tall, the flowers borne, slightly nodding, on erect stems, and pale scarlet with a narrow golden throat, 2 inches (5 cm) across and opening stiffly flat.

'**A. J. Hogan**', 1994, Hogan. *Crocosmia ×crocosmiiflora* × 'Lana de Savary'. Vigorous, with bright red, upward-facing, trumpet-shaped flowers; about 30 inches (75 cm) tall.

'**Amber**', 1998, Dunlop. *Crocosmia masoniorum.* A soft amber-colored form, about 3 feet (90 cm) tall.

'**Amberglow**', 1984, Browne. 'Jackanapes' × 'Solfatare'. Soft orange flower, opening flat, about 1½ inches (38 mm) across, with purple spots around the cream center. Flowering at 24–28 inches (60–70 cm) tall, above bronze foliage.

'**Amber Sun**', 2003, Dunlop. A soft, warm orange flower about 2 inches (6 cm) wide, with tepals that reflex readily in warm conditions; plant 2 feet (60 cm) tall, with bronze foliage.

'**America**', 1895, Krelage? Very free-flowering, wonderful warm orange-red in which there is a suspicion of an apricot shading. The individual flower is not large, but there is a sunny look about it that is most effective; a very bright flower (Gumbleton 1905).

'**Anglia**', 1903, Davison. Pale golden yellow suffused with a little spangled reddish coloring, near the tepal tips, internally.

'**Anneau d'Or**', golden ring, 1900, Lemoine. Enormous, perfectly formed, brilliant orange flowers with a broad chestnut brown ring around the yellow center and with reflexed tepals; a superb flower, 2 inches (5 cm) across, the tepals broad and of wonderful coloring, an intense orange-yellow, with a velvety purple base and the center a lighter yellow. The buds are dark red. The backs of the tepals show very little of the purplish shade but, like the upper surface, are orange-red (Gumbleton 1905).

'**Anniversary**', 2000, Lewis. *Crocosmia masoniorum* × *C. paniculata.* A derivative of 'Lucifer' with a noticeable yellow eye.

'**Apricot Queen**', 1924, Fitt. Flowers of medium size, of a rich golden apricot; height 30 inches (75 cm) (Barr 1928, Cowley 1929).

'**Apricot Queen**', 1995, Dunlop. *Crocosmia masoniorum.* Pale apricot flowers. Name re-used because the original *C. ×crocosmiiflora* cultivar lost to cultivation.

'**Arc en Ciel**', rainbow, 1894, Lemoine. Vivid yellow-ochre flowers are almost as big as those of 'Imperialis' but droop a little; they are embellished with light brown markings around the golden yellow center, making a striking contrast with the rest of the flower.

'**Aurantiaca**', orangish, pre-1905, breeder unknown. Flower spikes much branched, bearing rich, golden yellow flowers 2 inches (5 cm) across, widely bell-shaped and slightly drooping. The spikes are a trifle crowded with flowers (GBM 1905).

'**Aurea**', 1884, Lemoine. A tall plant, carrying numerous spikes of flowers that are broad, open, of a beautiful golden yellow and producing great effect.

'**Auréole**', halo, 1899, Lemoine. Magnificent large flowers with broad tepals, perfectly opened and displayed; yellow outside and gilded yellow inside, with a pale straw yellow center surrounded with a crimson-maroon ring.

'**Auréole**', halo, pre-1910, Gerbeaux. Sulphur yellow flowers with the tepals flushed red at the tips (Gerbeaux 1910).

'**Auricorn**', 1996, Dunlop. *Crocosmia masoniorum × C. paniculata.* Soft orange flowers typical *C. masoniorum,* attractively set off against the distinctly dark green pleated foliage; about 3 feet (90 cm) tall.

'**Auricule**', little finger, 1892, Lemoine. Dark yellow flowers with a purple circle around the center.

'**Auriol**', 2003, Dunlop. 'Rowallane Yellow' × *Crocosmia pearsei.* Compact, about 28 inches (70 cm) tall, with well-formed orange flowers of good substance, with a yellow eye and pointed tepals, facing forward. The flowers have a long flowering period, so almost the whole panicle is open at the same time.

'**Aurora**', 1919, Fitt. Very large flowers, pure orange-yellow; tall-grower (Fitt 1920). KAVB Award of Merit, 1927.

'**Aurore**', dawn, 1890, Lemoine. Tall, branched stems with numerous large flowers 2³/₁₆ inches (56 mm) in diameter. The perfectly regular tepals are broad, rounded, well spread out and a magnificent yellowish orange.

'**Australiens Gold**', Australian gold, pre-1907, Krelage? No description traced (Gerbeaux 1907).

'**Autumn Gold**', pre-1955, breeder unknown. The first named cultivar raised in South Africa, it has smallish, trumpet-shaped, golden yellow flowers on well-branched stems about 3 feet (90 cm) tall and dusky green foliage (Eliovson 1955: plate 142).

'**Baby Barnaby**', 1992, Lewis. Charming, small, with chrome yellow flowers with dark throat markings (Lewis 1994).

'**Babylon**', pre-1991, van Dijk. Extremely vigorous with flowers 2¹/₂ inches (6 cm) across, rich red, with a small but conspicuous dark maroon ring around the fine yellow eye; about 32 inches (81 cm) tall.

'**Beth Chatto**', 1993, Chatto. *Crocosmia masoniorum* × *C. pottsii*? Red flowers, with narrow pleated foliage but with a flower typical of *C. masoniorum*.

'**Bicolore**', bicolored, 1895, Welker. Dwarf plants with a short spike and average-sized flowers. Three of the rounded tepals are red-orange, alternating with the other three vermilion-red ones; an extraordinary plant (Grignan 1907).

'**Bouquet Parfait**', perfect bouquet, 1885, Lemoine. Flowers are very large and a vivid vermilion with dark yellow centers. The flowering stems lean in all directions. KAVB Award of Merit, 1885.

'**Bressingham Beacon**', 1975, Bloom. *Crocosmia masoniorum* × *C. paniculata*. Plants about 3 feet (90 cm) tall and slow to increase. The flowers are red, brightly flushed gold in the center and presented like those of *C. masoniorum* but showing a slight influence from *C. paniculata* in their individual form, and the plant's slowness to increase vegetatively.

'**Bressingham Blaze**', 1970, Bloom. *Crocosmia masoniorum* × *C. paniculata*. Dwarf plants about 2 feet (60 cm) tall with an unbranched spike and rich red flowers presented like *C. masoniorum* but the individual flowers narrower in form, showing the influence of *C. paniculata*.

'**Brightest and Best**', 1922, Fitt. Orange, red shaded, dwarf growing (van Scheepen 1963). KAVB Award of Merit, 1926, First Class Certificate, 1927.

'**Brilliant**', 1897, Lemoine. Short plants with round, widely open flowers with broad scarlet tepals marked brown at the base.

'**Burford Bronze**', 1984, Browne. Flowers, about $1^1/_2$ inches (38 mm) in diameter, are a warm but not vivid yellow, well presented above the bronze foliage; about 3 feet (90 cm) tall. It is best likened to a taller and more robust form of 'Solfatare'.

'**Buttercup**', 1995, Heemskerk. Rich butter yellow flowers about $1^3/_8$ inches (35 mm) across, opening flat; a robust grower. Raised by Jac M. van Dijk but renamed as plants of the original name (possibly 'Polo') were lost.

'**Cadenza**', 1996, Dunlop. *Crocosmia pottsii* × 'Grandiflora'. Self-sown seedling with bright orange, narrow, trumpet-shaped flowers with a pink tinge to the inside of the tepals and a yellow throat, flowering at 4 feet (1.2 m) tall, with an extremely long inflorescence that gradually elongates as the long succession of flowers open.

'**Cadmium Star**', 2003, Dunlop. 'Voyager' × *Crocosmia aurea*. About 3 feet (90 cm) tall with chrome yellow, star-shaped flowers of good substance, about $2^1/_2$ inches (6 cm) in diameter, well spaced along the stem.

'**California**', 1894, Krelage. Light golden yellow, a delightful flower, warm, but bright in coloring. The broad florets make up an almost circular bloom $1^1/_2$ inches (38 mm) wide. Its tinge of color deepens toward the center, where there is a suffusion of yellow, and within the throat there is an irregular purple ring; among the brightest of all (Gumbleton 1905, Krelage 1896). Award of Merit at Amsterdam.

'**Canari**', canary, pre-1906, Welker. Short plants with compact spikes of vivid canary yellow flowers (Grignan 1906).

'**Canary Bird**', 1990. Vigorous, yellow, similar to 'Norwich Canary' but with narrower leaves. First listed by Broadleigh Gardens, Somerset, England; probably of Dutch origin.

'**Cardinale**', 2003, Dunlop. Rich red, upward-facing, trumpet-shaped flowers; plant 30 inches (75 cm) tall, on purple stems with relatively short green foliage.

'**Cardinalis**', scarlet, pre-1906, Welker. Short plants with tightly clustered, cardinal red flowers with paler centers (Grignan 1906).

'**Carmencita**', 1922, Fitt. Seedling No. 8. Very deep crimson-scarlet, yellow-marked center, very free-flowering, increases rapidly (van Tubergen 1926).

'**Carmin Brilliant**', 1950, Pfitzer? Carmine-colored. Listed by Pfitzer nursery (1953). RHS Award of Garden Merit, 2002. See also 'Mrs Geoffrey Howard'.

'**Carminea**', pre-1912, breeder unknown. Carmine scarlet, very distinctive (Prichard 1912). This might be the forerunner of 'Carmin Brilliant' or the surviving plant renamed after the Second World War by the Pfitzer nursery.

'**Carnival**', 1998, Dunlop. *Crocosmia masoniorum* × ? Openly pollinated hybrid of uncertain parentage, resembling *C. masoniorum* in its bright red flowers, flushed yellow in the center; about 39 inches (1 m) tall.

'**Carola**', 1982, Zonneveld. Flowers large, *Capsicum* red with marigold orange midrib, flower tube cadmium-orange outside, tepals nasturtium red, anthers bright yellow (van Scheepen 1991).

'**Cascade**', 1999, Dunlop. *Crocosmia paniculata* × *C. masoniorum*. Robust, about 39 inches (1 m) tall, with multibranched, drooping inflorescence with pale orange, trumpet-shaped flowers.

'**Castle Ward Late**', pre-1895, Leichtlin. *Crocosmia paniculata* × *C. aurea*. One of five named cultivars of *C. ×crocosmoides* still in cultivation. Vigorous, leaves pleated.

'**Cavalier**', 1997, Mahir. Compact, 2 feet (60 cm) tall, the flowers light orange with a wide throat of pale yellow extending to almost half the length of the broad tepal, opening flat.

'**Cecile**', 1922, Fitt. Yellow-flowered cultivar of good size and substance, having the individual blooms nicely disposed on the spike. RHS Award of Merit, 1923.

'**Céres**', 1891, Lemoine. Medium-sized flowers are an apricot orange color, with broad tepals opening widely, facing upward and presented on tall, branched stems.

'**Challa**', pre-1991, van Dijk. Flowers large, with outer tepals brown-red outside, brownish red inside, the inner tepals Indian yellow,

nasturtium orange with a brownish red vein outside (van
Scheepen 1991).

'Chinatown', 1998, Dunlop. *Crocosmia masoniorum* × *C. paniculata*.
Bright red flowers with a conspicuous yellow eye and perianth
tube; flowers 4 feet (1.2 m) above pleated foliage.

'Chloris', pre-1904, breeder unknown. Flowers broad, bell-shaped, a
deep saffron yellow, and the spikes 10- to 12-branched, fully 2 feet
(60 cm) high, and flowering very freely, a display being maintained for 6 weeks (GBM 1904).

'Chrysis', golden, 1896, Welker. Vigorous plants with very large, open
flowers 2½ inches (6 cm) in diameter; the flowers are a brilliant
apricot yellow with a light yellow center (Grignan 1906).

'Citronella', 1917, Henley. 'Messidor' × 'Gerbe d'Or' (parentage from
J. E. Fitt notebooks). Robust grower 3–4 feet (90–120 cm) high,
bearing lovely flowers of citron yellow (Fitt 1920). The flowers are
of a soft cool yellow tone with two slight maroon flashes near the
eye (Cowley 1918). The name is currently wrongly applied to one
or several of the small, pale yellow-flowered plants to which a variety of other names have been attached: 'Golden Fleece', 'Golden
Sheaf', 'Honey Angels', 'Loweswater' and 'Mount Usher'. RHS
Awards of Merit were given following trials in the 1960s and 1980s
but to different plants, neither of which was correctly identified as
this cultivar.

'Coleton Fishacre', pre-1980, breeder unknown. Self-sown seedling.
'Coleton Fishacre' is perhaps the most appropriate name for a
plant that was discovered in the Coleton Fishacre Garden, Devon,
England, along with 'Lady Hamilton' and 'Solfatare'. Coleton
Fishacre was originally the home of the D'Oyly Carte family but is
now in the care of the National Trust. The cultivar has distinct
bronze foliage and smallish, soft orange, trumpet-shaped flowers
that face upward. It is quite vigorous and robust, flowering at
about 30 inches (75 cm) tall. A variety of names have been attached to it: 'Coleton Fishacre Solfatare', 'Rowden Bronze' and
'Dark Leaf Apricot'. It was wrongly identified by Gary Dunlop as
the Lemoine cultivar 'Gerbe d'Or', from an inadequate early description; it is now widely grown under the name 'Gerbe d'Or'.

'Coleton Fishacre Solfatare', synonym of 'Coleton Fishacre'.

'Columbus', 1999, Heemskerk. Robust and vigorous, with rich, golden yellow flowers about 1⁹/₁₆ inches (4 cm) across, with several small dark blotches near the center, opening from distinctive violet-colored bracts. Originally raised by Jac M. van Dijk.

'Colwall', 1985, Picton. Sturdy, robust and red-flowered, with bronze-tinted foliage; flowers about 2¹/₂ inches (6 cm) across.

'Comet', 1909, Davison. Combining the rich coloring of 'Westwick' with the great size of 'Prometheus' and the best of the crimson-stained varieties (Wallace 1913). 'Comet' may be descended from 'Prometheus' and has all the fine qualities of that glorious montbretia. Its color is a little deeper, and the dark center gives this flower an aspect of its own. The blooms are large and well formed (Arnott 1910a). KAVB Award of Merit, 1938.

'Comet' (Plate 14), pre-1980, breeder unknown. *Crocosmia masoniorum* × *C. aurea*. Flowers intermediate between those of the parents, large and yellow, bright and striking. with tepals centrally streaked red longitudinally. The flowers are sideward-facing from a *C. masoniorum* form of inflorescence. Plants can sometimes be difficult to establish initially. The stoloniferous habit is derived from both parents, and exaggerated to the extent that satellite shoots can first appear above ground as much as 18 inches (45 cm) away from the original planting. The origins of this hybrid are obscure. Fred Knutty, of Malahide Nursery, Dublin, Ireland, obtained the plant as *C. masoniorum* in the 1970s. The name 'Comet' was inappropriately attached to the plant by Gary Dunlop, based on early descriptions of George Davison's cultivar of that name. As *C. masoniorum* did not enter cultivation in the British Isles until the 1950s, that identification was incorrect. The name is, however, still valid, as it was applied to a hybrid of different parentage and was published (Dunlop 1999). Several other names have been applied by various nurseries: 'Fred Knutty's', 'Malahide', 'Malahide Castle' and 'Jim Reynolds's'. The latter name derives from plants being subsequently distributed by Jim Reynolds from his Butterstream Gardens, County Meath, Ireland.

'Congo', 1897, Lemoine. Very floriferous plants of average size with brilliant golden yellow outer tepals spotted with light brown at the base, and some tepal-like stamens. A very dark orange-red hybrid, bud and open flower being of this very rich shading, reminding one of 'America' (Gumbleton 1905).

'Constance', 1993, van Dijk. Flowers large, marigold orange, outer tepals nasturtium orange outside, nasturtium red inside, inner tepals marigold orange, center buttercup yellow.

'Copper King', 1920, Fitt. Very distinctive and handsome, with flowers of fair size, dark coppery red with gold luster on tepals; dark stem, late flowering, 30 inches (75 cm) tall (Barr 1924).

'Corona', 1997, Dunlop. Rich orange flower about 2 inches (5 cm) across, opening flat, with an irregularly flared dark ring around the center. It is robust and flowers at about 3 feet (90 cm) tall.

'Corso', 1986, van Dijk. *Crocosmia masoniorum* × *C. ×crocosmiiflora*. Flowers large, brown-red glowing fire red, center tangerine orange, throat pale orange (van Scheepen 1991).

'Corten', 1997, Dunlop. Plants about 30 inches (75 cm) tall with flowers almost 2 inches (5 cm) across, dull rusty orange with a yellow eye.

'Couronne d'Or', golden crown, 1897, Lemoine. Tall plants with large, rounded, golden yellow flowers spotted at the throat.

'Crimson King', date and breeder unknown. No description traced; probably an Earlham Hybrid (Wallace 1926).

'Crimson Spotted', see 'Maculata'.

'Croesus' (Plate 13), 1903, Pfitzer. Fine large flowers of a rich orange-yellow. Listed as new (in uppercase) by Daisy Hill Nursery (1906) but available only by the dozen.

'Culzean Peach', synonym of 'Culzean Pink'.

'Culzean Pink' (Plate 15), date and breeder unknown. *Crocosmia pottsii*, probably a self-sown seedling. Robust and distinctive color form, almost 43 inches (1.1 m) tall, grown for many years in the garden of Culzean (pronounced 'kill-ain') Castle, Scotland. The upper half of the perianth tube and tepals are tinged a pale pink.

'Custard Cream', 1985, Browne. Similar to 'Morning Light' but with slightly paler flowers. It requires good cultivation conditions, otherwise it quickly reduces significantly in height, producing much smaller flowers. Its normal height is about 3 feet (90 cm).

'**Cyclops**', 1919, Fitt. 'Pageant' × 'Lord Nelson' (parentage from J. E. Fitt notebooks). Yellow with a deep purple ring around the center (Fitt 1920).

'**Daisy Hill**', 1913, Smith. Handsome, large-flowered; orange-yellow with dark zone. First listed by Smith (1914, 1915); only available singly, most expensive.

'**Dancing Ballerina**', 1900, Regney. *Crocosmia aurea.* A good form, well within the natural range of the species. It was raised from wild-collected seed, about 1985, subsequently given a cultivar name and protected by Plant Breeder's Rights about 1998.

'**Dark Leaf Apricot**', synonym of 'Coleton Fishacre'.

'**David Fitt**', 2002, Fenwick. 'Zeal Giant' × 'Lucifer'. Inflorescence similar to a red form of *Crocosmia masoniorum* with a faint orange flush on the tepals, presented on a dark purple stem. Named after a son of J. E. Fitt.

'**Davisoni**', see 'George Davison'.

'**Debutante**', 1994, Browne. Small, upward-facing flowers that open orange internally but quickly fade to pale pink, with a yellow eye, contrasting well with the deeper orange-red exterior. The flowers are presented at about 32 inches (81 cm) high, just clear of the green foliage.

'**Devil's Advocate**', 1998, Fenwick. Deeper, somber red variant raised from seed of 'Lucifer'.

'**D. H. Houghton**', 1930s, Fitt. Deep orange flowers (also described as red) on a well-formed spike; strong grower (Simpson 1937).

'**Diadème**', tiara, 1897, Lemoine. Flowers are golden yellow with an orange tint, and a brown ring in the center.

'**Distinction**', 1898, Lemoine. Round flowers are brick orange in color, with very large golden yellow centers.

'**Dixter Flame**', about 1990, Lloyd. *Crocosmia masoniorum.* A rich, bright red form.

'**Doctor Hoog**', 1894, Krelage. Dazzling yellow, fine flower (Krelage 1896); discontinued by 1900. Award of Merit at Amsterdam.

'**Dr Marion Wood**', 1999, breeder unknown. *Crocosmia masoniorum* × *C. paniculata*. Large pinkish red flowers with the tepals suffused with soft orange.

'**Doctor Masters**', 1893, Krelage. Novelty (Krelage 1896). Award of Merit at Amsterdam.

'**Donegal**', 1993, Chatto. A form of the common montbretia from County Donegal, Ireland.

'**Dora**', 1933, Fitt? Light orange flowers (Prichard 1935).

'**Drap d'Or**', cloth of gold, 1887, Lemoine. Carries itself perfectly erect; the flowers are very large, with broad chrome yellow lobes—an eye-catching plant.

'**Duchess of Mecklenburg**', 1906, Krelage. Tall, with bright orange flowers (Hayward 1910).

'**Dusky Maiden**', 1989, Browne. Plants about 2 feet (60 cm) tall, with dull orange flowers, the color of Corten steel, rust-colored, blending subtlety with the dark bronze leaves. Its somber color scheme provides a contrast to the more usual vivid colors of the genus.

'**E. A. Bowles**', 1925, Fitt. 'Sunshine' × seedling. Flowers a lovely shade of rose-cardinal with a pale crimson zone and a yellow throat. RHS Award of Merit, 1926; KAVB Award of Merit, 1930.

'**Early Bird**', 1992, Mahir. *Crocosmia masoniorum* × ? Deep scarlet flowers with a ruby sheen and golden throat, of the form characteristic of *C. masoniorum* and presented on stems about 3 feet (90 cm) high, and with typical pleated foliage.

'**Eastern Promise**', 1993, Lewis. Flowers have a soft orange exterior and bright yellow interior, a combination displayed to good effect in a clump. Tall and elegant, flowers like those of 'Star of the East', only smaller, and distinctly bronze midribbed foliage (Lewis 1994).

'**Éclatant**', dazzling, 1894, Lemoine. Dwarf plant, flowers tilted but a brilliant, crimson, blood red, inset with golden yellow marks.

'**Eldorado**', city of gold, 1887, Lemoine. Dwarf plant, very floriferous, growing to 14 inches (35 cm) tall at most; flowers large, golden yellow—a very good cultivar. KAVB Award of Merit, 1887. The name 'Eldorado' was wrongly attached to the Earlham Hybrid 'E. A. Bowles' in the late 1980s.

'Elegance', 2003, Dunlop. *Crocosmia paniculata* × *C. masoniorum.* Plants 5 feet (1.5 m) tall, with sprays of middle orange flowers presented on long, well-spaced and elegantly curved panicles.

'Elegans', 1883, Lemoine. Reminiscent of the original *Crocosmia* ×*crocosmiiflora,* with brilliant yellow flowers internally; the exterior tube and buds are vermilion on a yellow base—a plant of great effect.

'Elegantissima', most elegant, pre-1895, Welker. No description traced (Journal de la Société Nationale d'Horticulture de France 1895: 544).

'Ellen', pre-1991, van Dijk. Flowers large, the outer tepals reddish brown outside, inner tepals maize yellow with reddish brown tip outside, inside Saturn red, center tangerine orange, throat pale orange (van Scheepen 1991).

'Elsie', pre-1925, Fitt. Flowers deep red with a yellow center, very effective, star-shaped, $4^3/_8$ inches (11 cm) across; about 3 feet (90 cm) tall (van Tubergen 1926).

'Emberglow', 1970, Bloom. *Crocosmia pottsii* × *C. paniculata.* Clusters of rich, sultry red flowers similar in form to those of *C. pottsii,* on stems 3 feet (90 cm) tall, with pleated foliage; one of the darkest red-flowered hybrids. It sets seed readily and fruiting stems are sold as cut flowers in the Netherlands.

'Embrasement', fire, pre-1910, Gerbeaux. No description traced (Gerbeaux 1910).

'Emily McKenzie' (Plate 15), 1951, McKenzie. Flowers $2^1/_2$–$2^3/_4$ inches (6–7 cm) wide and tepals $^3/_4$ inch (18 mm) across at their widest part, the color a shade of Orpiment Orange between 10/1 and 10 on the *Horticultural Colour Chart* (probably bright orange). Three of the tepals have a central blotch of Chrysanthemum Crimson (824 *HCC*) at the base, and the other three tepals have two opposite marginal blotches. The spikes grow 30 inches (75 cm) tall and are well furnished with flowers. It was claimed that the plant needs no protection whatsoever and could be left in the ground all year-round. RHS Award of Merit, 1954; KAVB Award of Merit, 1957, First Class Certificate, 1958.

'Ernest Davison', 1903, Davison. 'Germania' × 'Prometheus'. Vigorous and sturdy-growing plants with many branched spikes of deep orange, flat flowers, 3 inches (7.5 cm) in diameter, flushed red externally and suffused carmine on the inner tepals around a large golden center.

'Étincelant', sparkling, 1891, Lemoine. Straight, branching stems support flowers with wavy tepals. Flowers face outward and are blood scarlet with sulphur-colored centers surrounded with purple marking; the reverse is yellow and blood red, and the buds are blood red with a yellow base—a plant with a grand effect.

'Etna', 1893, Welker. Wide-open, large flowers, the inside lava red–orange, the outside vermilion-red (Journal de la Société Nationale d'Horticulture de France 1895: 544).

'Étoile de Feu', fire star, 1885, Lemoine. Flowers are larger than those of *Crocosmia* ×*crocosmiiflora* and open widely; they are blood red on the outside with a vermilion interior and light yellow center, a dazzling tint.

'Étoile d'Or', golden star, 1893, Lille. Listed by Lemoine in 1893 or 1894 without description; subsequently described as having yellow flowers (Lemoine 1895).

'Étoile Polaire', polar star, 1906, Gerbeaux? Large yellow flowers with the tepals tipped red and crimson-red on the outside (Gerbeaux 1907).

'Europa', Europe, 1904, Pfitzer. Medium-sized but very shapely flowers on erect stems, the color a most striking tone of crimson-scarlet. Listed as new (in uppercase) by Daisy Hill Nursery (1906) and only available singly.

'Excelsior', pre-1911, Polman? Pure yellow, gold-dusted, violet-like flower 3 inches (7.5 cm) across, of most perfect shape; a gem in every respect (Backhouse 1911).

'Exposition 1900', 1900, Gerbeaux? No description traced (Gerbeaux 1910).

'Fandango', 1996, Dunlop. *Crocosmia paniculata* × *C. masoniorum*. Fairly vigorous, free-flowering, with bright, pale orange flowers displayed at more than 39 inches (1 m) high, above the narrow pleated foliage.

'Fantaisie', fantasy, 1892, Lemoine. Very large, well-branched panicles on erect stems, supporting medium-sized symmetric flowers presented very full face, with a broad, light yellow center, and the tepals tipped a vivid vermilion, red on the reverse—a perfect version of 'Bouquet Parfait'.

'**Fern Hill**', 1995, breeder unknown. *Crocosmia masoniorum*. A form with soft orange flowers with a yellow eye, from the garden of Fernhill, Sandyford, County Dublin, Ireland, owned by Mrs Walker.

'**Festival Orange**', 2002, breeder unknown. Medium-sized orange flowers with a brown throat and dusky foliage.

'**Feu d'Artifice**', firework, 1891, Lemoine. Tall, floriferous, with straight, well-branched stems supporting very large flowers that are round, spread out and regular, with broad, brilliant cadmium yellow tepals tipped with red.

'**Feu Follet**', will o' the wisp, 1899, Lemoine. Dwarf floriferous plants with yellow-ochre flowers with maroon spots around the center, showing through to the outside.

'**Fiery Cross**', 1930, Fitt. Plants 42 inches (1.1 m) tall, erect, well branched; stems strong, stiff. Flowers very freely borne, $2^3/4$ inches (7 cm) across, flat, of good substance, orange-red with large, pale yellow zone at throat; a strong, sturdy grower. RHS Award of Merit, 1931.

'**Figaro**', 1900, Lemoine. Large, fully open, vivid orange flowers with a central broad, chestnut ring, facing upward from straight stems.

'**Firebird**', 1970, Bloom. *Crocosmia masoniorum* × *C.* ×*crocosmiiflora*? This cultivar has the appearance of *C. masoniorum* but with noticeably larger, rich red flowers; about 30 inches (75 cm) tall.

'**Firebrand**', 1985, Browne. 'Jackanapes' × 'Solfatare'. Red flowers with yellow centers, opening wide, about $1^1/4$ inches (32 mm) across, displayed at about 30 inches (75 cm) above green foliage.

'**Firefly**', 2002, Dunlop. *Crocosmia paniculata* × *C. pottsii*. Small, rich red flowers are well spaced on widely branched panicles; about 39 inches (1 m) tall.

'**Fireglow**', 1985, Browne. Orange flowers about $1^1/4$ inches (32 mm) across, with small purple-marked yellow centers; about 2 feet (60 cm) tall.

'**Fire Jumper**', 2001, Hinkley. Relatively large flowers possessing nicely rounded, overlapping tepals of golden yellow suffused with orange in the throat, opening from tangerine buds along sturdy stems as long as 2 feet (60 cm) (Hinkley 2003).

'Fire King', 1905, Krelage. Medium-sized flowers of dazzling deep crimson-scarlet; a very abundant bloomer and useful cultivar. See also 'Jackanapes'.

'Fire Sprite', 2001, Dunlop. *Crocosmia pottsii* × *C. ×crocosmiiflora*. Plants about 30 inches (75 cm) tall, with rich, red, trumpet-shaped flowers facing upward, along the dark stem.

'Fireworks', 2000, Lewis. *Crocosmia pottsii* × ? Bright red, yellow-eyed flowers on stems about 3 feet (1 m) tall.

'Flaire', 2002, Dunlop. *Crocosmia pearsei* × 'Rowallane Yellow'. Upward-sweeping, large, widely flaring trumpets of rich orange; the flower form is intermediate between the two parents. Flowers at about 39 inches (1 m) tall but is very slow to increase.

'Flamboyant', pre-1906, Welker. Floriferous plants with uniformly colored and sized red lead–orange flowers (Grignan 1906).

'Flamenco', pre-1975, Smith and Archibald. *Crocosmia masoniorum*. A rich red form 'selected from among several hundred seedlings' (Archibald 2000).

'Flamethrower', 1985, Browne. 'Jackanapes' × 'Solfatare'. Orange-red flowers about 1^1/$_2$ inches (38 mm) across, with yellow centers; 2 feet (60 cm) tall, with bronze foliage.

'Fleuve Jaune', yellow river, 1892, Lemoine. Very floriferous plants with erect flowers colored Naples yellow; an improvement on 'Solfatare'.

'Flore Pleno', doubled flower, 1892, Lemoine. A tall vigorous plant with broad, semireflexed leaves and straight panicles loaded with large, erect, perfectly double flowers, recalling double begonias in their appearance and mode of opening out. Their color is a brilliant orange-yellow. Of the first rank in every respect, forming magnificent specimens whether grown in pots or in beds, 'destined to become the source of a new fashionable race of double montbretias'. RHS First Class Certificate, 1893.

'Floribunda', many-flowered, 1892, Lille. Dwarf plant; flowers yellow and orange (Lemoine 1894).

'**Frans Hals**', 1913, Jan Roes. First listed 1916, with date, and claimed raising. Flower very large, widely opening, clear orange-yellow with a small brown ring; stems straight and strong. KAVB Award of Merit, 1913.

'**Fuego**', 2001, breeder unknown. *Crocosmia paniculata* × *C. masoniorum*. Appears to be very similar to 'Lucifer'.

'**Fugue**', 2003, Dunlop. *Crocosmia paniculata* × *C. masoniorum,* or a hybrid with the latter parent dominant. Very sturdy, upright growth habit, with large, light orange flowers on as many as six compact branches, arranged vertically on the stem and opening simultaneously, thus providing a very substantial appearance.

'**Fusilade**' (sic), 1995, Dunlop. *Crocosmia paniculata* × *C. masoniorum.* Short, robust, open-growing plants, that is, not compact but not exactly diffuse, about 27 inches (69 cm) tall, with rich red flowers flushed yellow in the center, and stocky pleated foliage.

'**George Davison**', 1900, Davison. 'Golden Sheaf' × (*Crocosmia* ×*crocosmiiflora* × 'Golden Sheaf'). Large, early, orange-yellow flowers; 8- to 10-branched, 3–4 feet (90–120 cm) high, bearing lovely pale orange-yellow flowers 3 inches (7.5 cm) across, widely expanded and tinted deeper orange externally; among the first to flower and strongly recommended as one of the most vigorous (Wallace 1913). RHS Award of Merit, 1902. See also 'Norwich Canary'.

'**George Henley**', 1908, Henley. 'Hereward' × 'King Edmund'. Flowers 3 inches (7.5 cm) across, bright chrome yellow tinged with red in the eye, of good form and substance. Stems 30 inches (75 cm) high, scarcely branched, holding flowers erect. Buds deep orange just before opening. RHS Award of Merit, 1909.

'**Gerbe d'Or**' (Plate 15), sheaf of gold, 1885, Lemoine. Very floriferous dwarf plants of good habit. The flowers are a little tubular but magnificent golden yellow.

'**Germania**', 1899, Pfitzer. 'Étoile de Feu' × 'Imperialis'. Growing 42 inches (1.1 m) high and with remarkably large, expanded flowers of a brilliant orange-red with a deep red center. RHS Award of Merit, 1901.

'**Gigantea**', 1910, Roes & Goemans. Listed 1911; large, yellow.

'Globe d'Or', golden sphere, 1898, Lemoine. A perfection of 'Flore Pleno'; the stems are dressed with full golden yellow flowers.

'Gloire de la Celle St Cloud', glory of St Cloud, pre-1906, Welker. Tall plants with broadly segmented pale apricot flowers, the three outer tepals marked with amaranth red (Grignan 1906).

'Gloire de Vitry', glory of Vitry, pre-1895, Welker. No description traced (Journal de la Société Nationale d'Horticulture de France 1895: 544).

'Gloria', 1933, Fitt? Medium-sized, clear yellow flowers, deepening at the margins (Simpson 1937).

'Golden Fleece', 1993, breeder unknown. Name applied to a freely branching form with small, pale yellow, trumpet-shaped flowers, similar to what is widely, but wrongly, in commerce as 'Citronella'. It is similar to 'Gerbe d'Or', also called 'Golden Sheaf', Honey Angels' and 'Loweswater'. 'Sulphurea' is similar but taller and more robust. There are at least six similar clones in cultivation under differing names. The name, in French, was originally used by Lemoine for a large-flowered yellow cultivar, 'Toison d'Or'.

'Golden Girl', 1997, Mahir. Plants about 3 feet (90 cm) tall and vigorous, with *Watsonia*-like flowers 2 inches (5 cm) across, a clear apricot color with carmine dots at the throat; tepals broad, pointed and recurved.

'Golden Glory', van Nieuwkoop. 1954 Pure apricot yellow flowers flushed darker inside toward a red-brown ring around the throat (Thomas 1982). KAVB Award of Merit, 1955.

'Golden Ray', 1912, Hayward. Large, pure golden yellow flowers, opening flat, nicely spaced on branching stems (Hayward 1914).

'Golden Sheaf', see 'Gerbe d'Or', but the English name is misapplied.

'Goldfinch', 1918, Henley. Two upper tepals bright red outside, and inside golden yellow with purple spots around the base of the tepals.

'Gold Sprite', 2000, Dunlop. *Crocosmia* ×*crocosmiiflora* × ? About 2 feet (60 cm) tall, with bright, chrome yellow, trumpet-shaped flowers facing upward.

'Goldstrom', gold stream, 1911, Pfitzer. The color of the flowers is a golden gamboge yellow of brilliant effect, increased by the purity of the shade, without any marking in the throat. The almost erect, well-branched stems, showing each flower to full advantage, and the profusion of spikes gives an effect of a mass of flowers without any trace of foliage. Listed as new by Pfitzer nursery (1913).

'**Gonnerbloem**', pre-1911, Polman? Fine orange flower, very free-blooming (Backhouse 1911).

'**Grandiflora**', large-flowered, 1887, Lemoine. *Crocosmia pottsii.* Very large flowers in erect spikes, the exterior reddish scarlet and the interior orange-yellow; very effective (Bull 1891). See also 'Superba'.

'**Grandiflora Flore Pleno**', large, double-flowered, 1894, Gerbeaux. *Crocosmia pottsii.* Small, orange-yellow, fully double flowers (Lemoine 1899).

'**Grandiflora Superba**', large-flowered, superb, 1892, Lille. First listed in 1894 (Lemoine 1894). Producing graceful spikes of glowing vermilion flowers, shading to golden yellow; height 2 feet (60 cm) (Barr 1930).

'**Grand Moulin**', great mill, pre-1906, Welker. Star-shaped flowers with long narrow yellow tepals tipped with carmine-red (Grignan 1906).

'**Grenade**', 1903, Lemoine. Long, branching stems covered with reasonably sized double or full flowers with overlapping scarlet tepals and a gold center.

'**Grenadier**', 1926, Fitt. Large star-shaped flowers of vivid orange-scarlet with crimson flush and golden sheen, with large, golden center, slightly stippled and blotched; height 30 inches (75 cm) (Barr 1932).

'**Guardsman**', 1934, Fitt. A good cultivar, producing medium-sized flowers of a rich vermilion with a slight suffusion of rose-crimson, the center golden, with crimson blotches (Barr 1936).

'**Hades**', 1926, Fitt. Medium-sized, star-shaped flowers of a bright vermilion-scarlet, shot with gold, with a gold throat and small blotches of crimson at the eye; height 30 inches (75 cm). Very showy, one of the brightest colors raised at the time (Barr 1932).

'**Halo**', 1917, Henley. 'Pageant' × 'Lord Nelson' (parentage from J. E.

Fitt notebooks). Pale orange-yellow flower with a large dark ring around the center (Cowley 1918).

'**Harvest Sun**', 2003, Dunlop. Bright yellow flowers about 2 inches (6 cm) wide, with an irregular orange-brown zone in the three lower tepals; plant about 30 inches (75 cm) tall.

'**Heatherselt**', a name applied to an old cultivar, identity uncertain, with medium-sized orange flowers that Jim Reynolds of Butterstream Gardens, County Meath, Ireland, supplied to the RHS trial of *Crocosmia* in the late 1980s.

'**Henri Welker**', pre-1895, Welker. Early-flowering, vigorous plants with broad foliage; the large flowers are copper-vermilion with purple markings in the center (Journal de la Société Nationale d'Horticulture de France 1895: 544).

'**Henry VIII**', 1917, Henley. 'George Henley' × 'Pageant' (parentage from J. E. Fitt notebooks). Flowers 4–5 inches (10–12.5 cm) in diameter, bright orange with dark markings (Fitt 1920). Largest-flowered of the Earlham Hybrids, it has good, showy orange flowers but its size means it lacks the elegance of the dainty 'Queen Charlotte' or the lovely 'Queen of Spain' (Cowley 1918). KAVB Award of Merit, 1926.

'**Herbert Perry**', 1932, Fitt. 'Lord Lambourne' × Seedling 1/28. Flowers $3^1/_2$ inches (9 cm) across, deep orange-scarlet, throat yellow. RHS Award of Merit, 1933.

'**Hereward**', 1903, Davison. 'Germania' × ? Pale or clear golden orange flowers, widely expanded and recurved as in *Lilium martagon,* reddish on the outside; very similar to 'George Davison' but flowering a month later. RHS Award of Merit, 1908.

'**Hertogin Jan Albert van Mecklenburg**', Dutch translation of 'Herzogin Johann Albrecht zu Mecklenburg'; used by Krelage (1902).

'**Herzogin Johann Albrecht zu Mecklenburg**', Duchess of Mecklenburg, 1899, Krelage. Large, dazzling orange *Crocosmia aurea*-like flowers. Called 'Hertogin Jan Albert van Mecklenburg' by Krelage (1902) but this original form used in the 1904 catalogue.

'**Highlight**', 1998, Saunders. *Crocosmia masoniorum* × *C. paniculata.* Attractive, with pale orange flowers closely resembling those of *C. masoniorum* but with yellow centers. It was supplied to Bridgemere Nurseries under a wrong cultivar name by a Dutch supplier and was given a suitable new name before being put on sale (C. Saunders, pers. comm. 1999).

'**Hill House**', 2001, Hubbard. *Crocosmia masoniorum* × *C.* ×*crocosmi- iflora*? Rich, red-flowered, quite similar to but an improvement on 'Firebird', with large, radially symmetric flowers.

'**His Majesty**' (Plate 14), 1918, Henley. 'Lemon Queen' × 'Queen Boadicea' (parentage from J. E. Fitt notebooks). Described as a 'glorious', free-flowering, large cultivar and 'by far the finest' Earlham Hybrid that appeared to that date. The center of the flower is clear yellow and shades into brilliant crimson-scarlet, the latter color deepest at the tips of the broad tepals. RHS First Class Certificate, 1919; KAVB Award of Merit, 1925, First Class Certificate, 1926.

'**H. Marshall**', pre-1936, Fitt? Intense vermilion flowers stained rose, very large (Prichard 1936).

'**Honey Angels**', date and breeder unknown. Small-flowered, yellow, with trumpet-shaped flowers, best likened to a yellow-flowered *Crocosmia pottsii*. See also 'Golden Fleece'.

'**Hon. Mrs Edwin Montague**', 1927, Fitt. 'Nimbus' × 'His Majesty'. Flowers very large and deep orange with a golden yellow zone around the center, lightly blotched with crimson. RHS Award of Merit, 1928, highly commended after trial, 1939.

'**Imperialis**', 1885, Leichtlin. A large form of *Crocosmia aurea* distributed 1888 under the name 'Macrantha', but the name was changed to 'Imperialis' at the suggestion of Victor Lemoine. RHS Award of Merit, 1892, First Class Certificate, 1894.

'**Improved**', 1930s, Slinger. *Crocosmia pottsii*. Very deep red (Slieve Donard catalogue numbered 1935/6 but not dated). See also 'Superba'.

'**Inca**', 1934, Fitt. Producing fine, well-branched spikes of bloom with medium-sized, stiff, flat, open flowers of a star shape with blunted tepals, bright orange-red shading to gold at the base, the center surrounded with heavy crimson-maroon blotches (Barr 1936).

'**Incandescent**', 1889, Lemoine. Enormous, fully open, orange-scarlet flowers with yellow centers, held on tall, branched stems. This is a

very distinct and pretty montbretia with a widely expanded flower
and broad tepals. It is lit up with color and seems coruscated like
Nerine, scintillating with golden orange coloring, touched with
red toward the apex of the broad firm tepals. At the base there are
a number of dark almost black spots. The back of the flower is
flushed with orange-red; the buds are dark red (Gumbleton
1905).

'Incendie', fire, 1887, Lemoine. Flowers are a brilliant red with
purple flecks in the center; they are held erectly on tall straight
stems—the most dazzling and beautiful of all the highly colored
montbretias.

'Indian Chief', 1926, Fitt. Fine, broad-tepaled flowers of a clear glistening orange, the center suffused crimson and blotched maroon,
with a clear golden throat; reverse of the tepals shaded orange-
scarlet, a glowing color. A vigorous grower; height about 39 inches
(1 m) (Barr 1932).

'Irish Dawn', 1999, Harding. Medium-sized, chrome yellow flowers,
well presented on stems 2 feet (60 cm) tall, flowering early.

'Irish Flame', 1999, Harding. Rich, flame red flowers, compact and
free-flowering, about 30 inches (75 cm) tall.

'Irish Sunset', 2000, Harding. Medium-sized yellow flowers, flushed
red on the outside.

'Jackanapes' (Plate 14), 1970, Bloom. Dwarf, 2 feet (60 cm) tall,
small-flowered. The insides of the tepals are alternately orange-
yellow and red; externally they are alternately light and dark red.
Probably the true 'Bicolore' raised by Welker in 1895, Bloom may
have obtained the plant in Northern Ireland in the late 1960s,
where it survives in old gardens under the wrong name 'Fire King'.

'James Coey', 1919, Fitt. Flowers of this handsome cultivar measure
4 inches (10 cm) across and are of deep orange–vermilion–red
color. The tepals are shaded with yellow and blotched with crimson at the base. Robust grower with large, rich dark red flowers,
with a very distinctive pale orange center (Fitt 1920). RHS Award
of Merit, 1920; KAVB Award of Merit, 1927. See also 'Mrs Geoffrey
Howard'.

'James Cross', 1926, Fitt. Very large, flat, open flowers with broad rounded tepals of an intense bright golden apricot, the center blotched and suffused rose-crimson; height about 3 feet (90 cm) (Barr 1932).

'Jennine', 1996, Dunlop. Clear yellow self-sown seedling, very similar to 'Jenny Bloom' but slightly shorter; it was obtained as that cultivar. The flowers of 'Jenny Bloom' are held above the stem whereas those of 'Jennine' are slightly smaller and presented below the stem. It is likely that 'Jennine' was a seedling almost identical to 'Jenny Bloom' and mixed in with it but never noticed.

'Jenny Bloom', 1975, Bloom. *Crocosmia pottsii* × *C. paniculata?* but may be of more complex parentage. Long, slightly branched spikes of chrome yellow flowers face upward; plants about 3 feet (90 cm) tall and excellent as a cut flower. Protected by Plant Breeder's Rights.

'Jessie', 1922, Fitt. 'Sunshine' × 'Pocahontas'. Medium-sized flowers of an orange shade tinged with pink while the center is yellow. RHS Award of Merit, 1923. At the time judged quite a new break in coloring of montbretias, of a very pleasing shade of shrimp pink with a soft yellow center, and purple spotted; very free-flowering (Fitt 1924).

'Jessie van Dyke', pre-1990, breeder unknown. Soft yellow flowers about 2 inches (5 cm) across, with several maroon blotches near the center on the lower tepals. Plants are about 30 inches (75 cm) tall, with narrow foliage, and increase quite vigorously. Probably of Dutch origin.

'Jewel', 1992, Mahir. Purple stems 2 feet (60 cm) tall, bearing outward-facing dusky red flowers 2 inches (5 cm) across and heavily flushed purple on the lowest tepals; tepals broad, rounded and opening flat from purple bracts.

'Joan of Arc', 1919, Fitt. 'Queen Adelaide' × 'George Henley' (parentage from J. E. Fitt notebooks). Large, clear apricot-yellow flowers faintly marked with red, measuring 3½ inches (9 cm) across. RHS Award of Merit, 1920.

'John Allan Fitt', 1932, Fitt. 'His Majesty' × 'Lord Lambourne'. Flowers 3 inches (7.5 cm) across, light orange-scarlet, throat yellow. RHS Award of Merit, 1933.

'**John Boots**', 2000, Heemskerk. Yellow flowers, opening flat, about 1^9/$_{16}$ inches (4 cm) across, with a slightly paler center. Probably originally raised by Paul Boots.

'**Judith**', pre-1991, van Dijk. Flowers large, outer tepals orpiment orange, inner tepals saffron yellow veined orange-yellow outside, inside orpiment yellow with a dark red spot near the base (van Scheepen 1991).

'**Judith Montague**', pre-1936, Fitt. 'Queen Mary' × 'Rose Queen'. Very free-flowering, the flowers close together on the stem, very large, yellow stained orange, fading to rose-cream (Prichard 1936). RHS Award of Merit, 1933.

'**Jupiter**', 1993, van Dijk? *Crocosmia masoniorum* × *C.* ×*crocosmiiflora*. Distinctive, with soft orange flowers attractively streaked from the center with a darker shade.

'**Kapoor**', 2002, Dunlop. Orange flowers, opening flat, about 2^1/$_5$ inches (5.6 cm) across, gradually suffusing to a much deeper color in the center; flowers at 30 inches (75 cm) tall and slow to increase.

'**Kathleen**', pre-1923, Fitt. Flowers very large and of great substance with broad apricot-scarlet tepals, the center primrose with conspicuous blotches of maroon, externally golden apricot with crimson markings (Barr 1925). Apricot-scarlet, primrose center with maroon blotches, reverse of the tepals golden apricot with crimson markings (Fitt 1928). KAVB Award of Merit, 1926.

'**Kiaora**', 1993, Dunlop. *Crocosmia masoniorum*. A soft orange form with a large yellow eye, with distinct dark blotches near the center, flowering at about 3 feet (90 cm) tall.

'**Kiatschou**', 1901, Pfitzer? Deep golden yellow flowers; very free. Listed as new (in lowercase) by Daisy Hill Nursery (1906) but available only by the dozen.

'**King Edmund**', 1904, Davison. Sturdy plants 3–4 feet (90–120 cm) high, stems freely branched, bearing close spikes of rich golden yellow flowers, deeper colored externally, opening flat, fully 3 inches (7.5 cm) across and marked with six to eight chocolate spots at the throat; vigorous. RHS Award of Merit, 1907.

'**Klondyke**', pre-1916, breeder unknown. Large flowers of rich golden yellow, with bronzy zone around the center of the tepals (Daisy Hill Nursery 1916).

'**Koh-i-noor**', mountain of light, pre-1906, Polman. Flowers 2$^9/_{16}$ inches (6.5 cm) in diameter, a clear orange-yellow with a paler yellow center and red markings at the base of two or three of the tepals, which are elegantly folded and embossed in two dimensions (Grignan 1906).

'**Lady Bangor**', synonym of 'Sulphurea'.

'**Lady Churchill**', 1926, Fitt. A choice cultivar, flowers almost a deep cardinal red, with white-shaded center; freely branching habit, very effective when massed (Fitt 1928).

'**Lady Diana**', 1926, Fitt. Very pretty shade of peach and red, center yellow (Fitt 1928).

'**Lady Gwen**', about 1930, Hill. 'His Majesty' × 'Indian Chief'. Scarlet flowers of medium size and with an orange-yellow throat. RHS Award of Merit, 1932.

'**Lady Hamilton**' (Plate 13), 1906, Davison. A wonderfully neat cultivar, growing 3$^1/_2$ feet (1 m) high, stems erectly held and studded with yellow flowers; exactly like a *Watsonia* in shape, flower and habit. A ray of rosy orange suffuses the center of the flower when it first opens, and the general coloring changes to apricot with age. RHS Award of Merit, 1907; KAVB Award of Merit, 1910.

'**Lady Oxford**', pre-1936, Fitt? Reddish orange flowers shaded offrose, the reverse slightly spotted yellow (Prichard 1936).

'**Lady Wilson**', 1926, Fitt. 'Henry VIII' × 'His Majesty'. Large, bright yellow with a very effective sheen of orange. RHS Award of Merit, 1927, and 1939 after trial; KAVB Award of Merit, 1930. See also 'Norwich Canary'.

'**Lambrook Gold**', 1998, breeder unknown. Smallish, golden yellow-flowered, of Dutch origin, quite similar to the plant often wrongly sold as 'Lady Wilson' but of deeper color. Named by Alan Street of Avon Bulbs, Somerset, England.

'**Lana de Savary**', 1993, Hogan. *Crocosmia* ×*crocosmiiflora* × 'Lucifer'. Robust, rich, bright red, free-flowering plants about 3 feet (90 cm) tall; the influence of 'Lucifer' is not very obvious.

'**Late Lucifer**', Lloyd. Similar to 'Lucifer' but later-flowering, with rich, dusky red flowers.

'**Leaping Salmon**', 1997, Mahir. *Crocosmia masoniorum* × ? Exceptionally long-flowering and vigorous, the flowers pale salmon-scarlet, presented on stems about 3 feet (90 cm) high, with pleated foliage.

'**Lemon Queen**', 1917, Henley. 'Gerbe d'Or' × 'Messidor' (parentage from J. E. Fitt notebooks). Chrome yellow with pale center, opening from deep orange buds (Fitt 1920). Similar to 'Citronella' (Cowley 1918).

'**Lena**', pre-1931, Fitt. Flowers orange-apricot with crimson blotches (ALW 1931).

'**Le Pactole**', gold mine, 1903, Lemoine. Tall, straight, well-branched stems with large brilliant golden yellow flowers presenting themselves upward. The excellent habit of this introduction makes it an acquisition of great merit. A beautifully colored flower 1$\frac{1}{2}$ inches (38 mm) across. Its charm is in the pure, satiny, almost apricot shade and the suspicion of crimson purple at the base of the three lower tepals. It is a firm flower, very free. The buds are orange-red; this shade is seen on the back of the expanded tepals (Gumbleton 1905).

'**Le Vainqueur**', the victor, 1897, Gerbeaux. Somber red flowers with a yellow throat marked with brown (Lemoine 1900).

'**Lord Lambourne**', 1925, Fitt. 'Queen Charlotte' × 'Pocahontas'. Handsome flowers very large, orange-scarlet and with irregular patches of dark crimson at the base of the tepals and orange-yellow throats. RHS Award of Merit, 1926.

'**Lord Nelson**', 1906, Davison. Deepest in color of all known montbretias at the time; growing about 3 feet (1 m) high and with very dark purple stems, much branched to form a fan-shaped inflorescence. Flowers 2$\frac{1}{2}$ inches (6 cm) across, deep orange-scarlet with a yellow eye, crimsoned externally, the tepals large and of unusually good finish. RHS Award of Merit, 1907.

'**Lothario**', 1900, Lemoine. Large golden yellow flowers, with broad tepals with small bronze spots at their base; they are presented

fully open and facing upward on straight, well-branched stems—
wonderfully bright, a brilliant apricot yellow and orange-red.
There are more distinct forms, but it is certainly a very cheerful
garden flower. It is free-flowering, and the buds have an orange-
red tinge (Gumbleton 1905).

'**Loweswater**', 2000, breeder unknown. Name given to one of the
forms of the small, pale yellow-flowered crocosmias with trumpet-shaped flowers. It was found in the garden of a deserted cottage in the Lake District in the north of England and named after
the adjoining lake. See also 'Golden Fleece' and 'Honey Angels'.

'**Lucifer**' (Plate 13), 1969, Bloom. *Crocosmia masoniorum* × *C. paniculata*. Rich red, robust hybrid flowers at more than 4 feet (1.2 m)
tall early in the season and is intermediate in form between its
parents. It comes true from seed and can self-seed. RHS Certificate of Preliminary Commendation, 1977; Award of Garden
Merit, 1993.

'**Lustre**', chandelier, 1895, Lemoine. Medium-sized yellow-orange
flowers lightly marked in the center and held on straight, upright
stems.

'**Lutea**', deep yellow, 1996, breeder unknown. Name attached to a
cultivar of Dutch origin, similar to the plants sold successively
under the names 'Lady Wilson', 'Norwich Canary' and, more recently, 'George Davison'. Its flowers are a slightly more orange
shade of golden yellow and about 1 inch (25 mm) across; plants
about 2 feet (60 cm) tall.

'**Lutea Praecox**', early-flowering deep yellow, 1903, Welker. Flowers
are a beautiful apricot yellow, held on very straight stems (Journal
de la Société Nationale d'Horticulture de France 1903: 529).

'**Lutetia**', pre-1906, breeder unknown. Bright red and deep yellow
flowers (Daisy Hill Nursery 1906).

'**Macrantha**', see 'Imperialis'.

'**Macrophylla**', large-leaved, 1907, Roes & Goemans. Listed 1908;
with broad, large leaves.

'**Maculata**' (Plate 5), 1886, O'Brien. *Crocosmia aurea* subsp. *aurea*.
Robust plants, to 3 feet (90 cm) high, with nodding orange flowers, the tepals bearing brown blotches at the base, the blotches

darker colored when grown in warm conditions. RHS First Class Certificate (under a wrong name, 'Crimson Spotted'), 1888, Award of Merit (though as a cultivar of *C. ×crocosmiiflora*, incorrectly), 1890.

'Major', 1899, Leichtlin. *Crocosmia paniculata*. A larger and later-flowering form of the species with bright red and yellow flowers.

'Mandarin', 1997, Dunlop. *Crocosmia paniculata × C. masoniorum*. Robust upright plants with largish, bright orange, trumpet-shaped flowers with a fine yellow eye.

'Marcotijn', 1982, Zonneveld. Flowers large, brown red with orange glow (van Scheepen 1991). The plant in cultivation under this name is a very vigorous hybrid, *Crocosmia masoniorum × C. pottsii*. It has soft orange *C. masoniorum*-type flowers with a broad brown ring around the yellow eye. More recently, the same plant has also been sold as both *C. masoniorum* and 'Sonate'.

'Margaret Rose', pre-1939, Fitt? Flowers fire red with yellow center, edged with purple (van Scheepen 1963). KAVB Award of Merit, 1957.

'Marjorie', 1918, Henley. Alike to 'Queen Mary' and 'Una' but with slightly different markings, and the tepals opening perfectly flat (Fitt 1920).

'Marjorie', 1926, Fitt. Large, open flowers, golden-orange shading to canary yellow, with crimson blotches around the eye; reverse of the tepals flushed orange-scarlet, very beautiful. Height 30 inches (75 cm) (Barr 1932).

'Mars', 1993, van Dijk? *Crocosmia masoniorum × C. paniculata*. Dwarf, red-flowered, with a compact inflorescence about 2 feet (60 cm) tall. Listed as new by van Dijk (1995).

'Martagon', 1897, Lemoine. Short plants with multiple spikes carrying very large flowers with reflexed tepals, like those of *Lilium martagon*. Three of the tepals are orange-yellow; the other three are bright dark red.

'Marthe Billard', pre-1895, Welker. No description traced (*Journal de la Société Nationale d'Horticulture de France* 1895: 544).

'Mephistopheles' (Plate 15), 1925, Fitt. Brightest-colored montbretia raised at the time, with broad, rounded tepals of a vivid flame scarlet shading to gold at the center, with crimson-maroon markings at the eye, the reverse of the tepals orange-scarlet and flame red, very beautiful; height 30 inches (75 cm) (Barr 1929).

'**Merryman**', synonym of 'Venus'.

'**Messidor**', 1899, Lemoine. Tall, straight, well-branched stems with flowers of Naples yellow and sulphur yellow tips fading to a very pale straw color in the center.

'**Météore**', meteor, about 1887, Lemoine. Flowers are very large, with compact, broad and rounded tepals; the inside face is dark brick yellow, the exterior blood red.

'**Michelle**', van Dijk. No description or date traced.

'**Minke**', pre-1991, van Dijk. Flowers large, outside saffron yellow, tipped marigold orange, inside orpiment yellow, center chrome yellow with tomato red blotches (van Scheepen 1991).

'**Minotaur**', 1994, Dunlop. *Crocosmia masoniorum* × ? Rich, deep orange-flowered, close to *C. masoniorum* but the inflorescence with two symmetric side branches that open simultaneously with the central flower spike.

'**Mr Bedford's**', synonym of 'Croesus', The name 'Mr Bedford's' applied by the late David Shackleton to a plant that he grew in his garden, Beech Park, southwest of Dublin, Ireland. It denotes that it was obtained from the garden at Straffen, County Kildare, where Mr Bedford had been an earlier head gardener.

'**Mistral**', 1998, Heemskerk. *Crocosmia masoniorum* × *C. paniculata.* Red-flowered, close in form to that of *C. masoniorum,* flowering at about 3 feet (90 cm). Originally raised by Langedijk nursery, the Netherlands.

'**Mrs Geoffrey Howard**' (Plate 15), pre-1935, breeder unknown. Listed by the Slieve Donard Nursery in Ireland from the mid-1930s for several years, and also in the 1950 by Daisy Hill Nursery. Similar to 'James Coey', both names appearing, with similar descriptions, in one undated Slieve Donard catalogue, number 16, about 1935. It is possible that it is an Earlham Hybrid; there is, however, no supporting documentary evidence, though J. E. Fitt was by far the most active breeder in the 1930s. The plant bearing the name 'Mrs Geoffrey Howard' is identical to an illustration of an early 1950s' illustration of 'James Coey' (Dix 1957); it thus may well be the true 'James Coey'. The old cultivar name 'James Coey' was, however, for many years used by the Dutch wholesale bulb trade for 'Carmin Brilliant'. This mistake has now been corrected, but the wrong use of 'James Coey' is still widespread in commerce and cultivation. In the circumstances, it is probably best that the name 'Mrs Geoffrey Howard' be maintained.

'Mrs H. J. Jones', 1919, Fitt. Medium-sized, well-shaped flower, maize yellow in the center, deepening to orange at the tip; the eye bearing distinct purple lines. Flowers externally carmine, deepening to vermilion with age, the contrasts in coloring very effective (Fitt 1920).

'Mrs Morrison's', a synonym, along with 'Mrs Thompson's', both applied to same plant that David Shackleton and Helen Dillon obtained from a Mrs Thompson of Rostrevor, County Down, Northern Ireland, who got it from a Mrs Morrison, who almost certainly obtained it locally from the Daisy Hill Nursery, Newry (H. Dillon, pers. comm. 1995).

'Mrs Stanley Baldwin', pre-1939, Fitt. Very handsome and exceptionally floriferous. The individual flowers are large and a brilliant orange-crimson, very effective (Albyn 1939).

'Mrs Thompson's', synonym of 'Mrs Geoffrey Howard'. See comment under 'Mrs Morrison's'.

'Mrs Warren Hulme', Fitt. No description or date traced.

'Moira Reed', 1997, breeder unknown. *Crocosmia masoniorum.* A form found in the garden of Moira Reed in Devon, with soft red flowers flushed with orange, and a conspicuous yellow throat.

'Moisson Dorée', golden harvest, 1907, Lemoine. Dwarf plants with broadly segmented flowers of a tawny golden yellow with a light chestnut ring around the sulphur yellow center.

'M. Jacqueau', pre-1910, Welker. Immense red and yellow flowers (Gerbeaux 1910).

'Moongold', 1933, Fitt. Very large, broad-tepaled, star-shaped flowers glistening rich yellow with clear cream center, reverse of the tepals and buds flushed apricot (Barr 1935).

'Moonlight', 1933, Fitt? Large, self-sown seedling with clear yellow flowers (Simpson 1937).

'Morgenlicht', morning light, 1894, Pfitzer. Listed at first without de-

scription but subsequently described as having dark yellow flowers, spotted brown (Lemoine 1896).

'**Morgenroth**', red dawn, 1895, Pfitzer. No description (Lemoine
1897).

'**Morning Light**', 1985, Browne. Clear yellow flowers about 1$\frac{1}{2}$
inches (38 mm) across, effectively presented on well-branched
stems about 30 inches (75 cm) tall, clear of the foliage. Similar to
'Custard Cream' but a more robust plant.

'**Mount Stewart**', synonym of 'Jessie'.

'**Mount Usher**', 1990, breeder unknown. Small, pale yellow-flowered,
named after the William Robinson-style garden south of Dublin,
Ireland, famous for its many rare and tender plants, where it survived and was rediscovered. It is similar to 'Golden Fleece'.

'**Nankin**', 1893, Welker. Quite large, open flowers, ochre-yellow inside (Journal de la Société Nationale d'Horticulture de France
1895: 544).

'**Newry Seedling**', 1912, Smith. Rich yellow with dark ring around
center; bold, flat flower. First listed by Smith (1914–1915); available by the dozen, expensive.

'**Nicolas Maas**', 1918, Jan Roes. Tall, strong-growing plants with clear
yellow flowers. Listed without description in 1920 but with description 1923; listed in 1926 as 'Nic. Maes'.

'**Nigricans**', blackish, 1914, breeder unknown. Light yellow with
bronze-green leaves (van Tubergen 1915); possibly 'Solfatare'
or a plant raised from it.

'**Nimbus**' (Plate 14), 1918, Henley. 'George Henley' × 'Pageant'
(parentage from J. E. Fitt notebooks). One of the most distinctive
of the Earlham Hybrids. The flowers are copper and gold, with a
distinct crimson ring around the center. RHS Award of Merit,
1918.

'**Norvic**', 1907, Davison. Dwarf, compact habit, but freely branched,
with dark stems and yellow flowers, tinged outside and in the bud
with red; 2 feet (60 cm) tall and late flowering. RHS Award of
Merit, 1908.

'**Norwich Canary**', 1990, breeder unknown. Small, golden yellow-
flowered cultivar of Dutch origin, about 2 feet (60 cm) tall with
flowers about 1 inch (25 mm) in diameter. The wholesale nursery

of Howard & Coeij was reputed to be the first to use this name for the Dutch plant then wrongly sold as 'Lady Wilson', which is lost to cultivation and had a much larger flower. 'Norwich Canary' is currently sold as 'George Davison'.

'Oeil de Dragon', eye of the dragon, 1896, Welker. Vigorous plants with large, open, fiery red flowers (Lemoine 1899).

'Old Hat', synonym of 'Walberton Red'.

'Olympic Fire', 2000, Hinkley. Somber bronzed swords of foliage and numerous flaming orange flowers, throated and backed with rich yellow, flowering late in the season on stems as long as 30 inches (75 cm) (Hinkley 2003).

'Orangeade', 1992, Knuckley. *Crocosmia masoniorum × C. paniculata.* Vigorous, free-flowering plants about 30 inches (75 cm) tall, with largish, bright red, broad, trumpet-shaped flowers facing sideways and well spaced along the inflorescence; probably of Dutch origin.

'Orange Devil', 2001, Saunders. *Crocosmia masoniorum × C. paniculata.* Form typical of many hybrids of this parentage, with rich orange-red flowers with a prominent golden orange suffusion in the inside. It was apparently a Bressingham Gardens reject, inadvertently sold by mistake as 'Lucifer' when that cultivar was first released (C. Saunders, pers. comm. 2003).

'Orange Glow', 1925, Fitt. Large, open, star-like flowers measuring 2½ inches (6 cm) across, of a rich orange with a golden sheen and vermilion flush, refined and beautiful (Barr 1929).

'Orange Lucifer', about 2001, breeder unknown. *Crocosmia masoniorum × C. paniculata.* Flowers are more similar to those of *C. masoniorum*, light orange with darker markings around the center. Possibly of Dutch origin.

'Oriflamme', banner, 1897, Gerbeaux. Dark red flowers with a red-brown center speckled yellow (Lemoine 1900).

'Oriflamme', banner, 1897, Lemoine. Large flowers open flat, with broad, rounded, scarlet-orange tepals; with a large golden yellow center and brown throat, surrounded by a dark red ring.

'Orion', pre-1915, breeder unknown. Shimmering yellow (van Tubergen 1915).

'Pageant', 1908, Davison. Stems 42 inches (1.1 m) tall, much branched, flowers 2 inches (5 cm) in diameter, orange-chrome with a red eye and a faint ring of red a short distance from the eye. RHS Award of Merit, 1909.

'Pamela', 1919, Fitt. Remarkably perfect in shape, with well-rounded, overlapping tepals. Flowers rich orange-scarlet with a soft yellow eye. Lower tepals handsomely blotched crimson-purple at the base. Good branching habit (Fitt 1920).

'Pepper', 1997, Fenwick. Short, of Dutch origin, with small, pale cream flowers 1³/₁₆ inches (3 cm) across, heavily flecked with orange.

'Peter', 1923, Fitt. Very free-flowering, making a wonderful display when massed. Flowers yellow with crimson tips, center shaded white; reaching a height of 4 feet (1.2 m) when well grown.

'Phare', lighthouse, 1886, Lemoine. Flowers are presented full face on densely clumping erect stems 30 inches (80 cm) tall. The color is richer than that of *Crocosmia* ×*crocosmiiflora;* the interior of the flower is a brilliant red lead color with a yellow center, and the exterior is a very vivid red lead. The buds are blood red at the tips and yellow at the base.

'Phoebus', bright, shining, 1925, Jan Roes. Orange flowers presented on a graceful flowering axis.

'Phyllis', 1926, Fitt. Similar in color and habit to 'His Majesty' but the center a shade deeper, with yellow and purple spots (Fitt 1928).

'Plancheon', van Dijk. No description or date traced.

'Plaisir', pleasure, 1999, Heemskerk. Rich orange flowers, opening flat, about 2 inches (5 cm) across, with a cream center surrounded with a maroon halo, flowering at about 2 feet (60 cm) tall. 'Emily McKenzie' is almost certainly one of its parents, and it is likely to have been originally raised by Jac M. van Dijk.

'Pluie d'Or', golden rain, 1889, Lemoine. Extremely floriferous and forming magnificent clumps. The flowering stems carry numerous sheaves of medium-sized yellow-ochre flowers, a new tint in crocosmias in 1889.

'Pocahontas', 1918, Henley. 'Queen Adelaide' × 'Lord Nelson' (parentage from J. E. Fitt notebooks). Flowers of a reddish terracotta color with a pale orange zone in the middle; widely expanded, star-shaped and of good size though not as large as 'Star of the East'. Very dark bright red, a strong grower (Fitt 1920). RHS Award of Merit, 1921.

'**Polo**', possibly a synonym of 'Buttercup'.

'**Polo**', 1995, van Dijk. Orange; no other description traced.

'**Précoce**', precocious, pre-1895, Welker. Citron yellow flowers spotted at the tip of the tepals are supported on fine, elegant stems (Journal de la Société Nationale d'Horticulture de France 1895: 544).

'**Prima Donna**', 1997, Mahir. Stems 30 inches (75 cm) tall, bearing flowers about $1^1/_2$ inches (38 mm) across, facing outward and scarlet flushed purple around the narrow golden throat. Rounded tepals open flat; their reverse and the short perianth tube are scarlet flushed purple. The bracts are deep purple.

'**Primrose**', 1919, Fitt. 'George Henley' × 'Tangerine' (parentage from J. E. Fitt notebooks). Clear yellow self-sown seedling with orange buds (Fitt 1920).

'**Princess**', 1922, Fitt. 'George Henley' × 'Queen Alexandra'. Large orange-red flowers with a yellow zone in the interior, the center surrounded by crimson spots. Deep orange tepals, with center pale yellow, with very dark crimson on the reverse side; the cultivar greatly admired when seen in beds (Fitt 1924). RHS Award of Merit, 1923.

'**Princess**', 1969, van Dijk. Large flowers, fire red, center yellow with purple-red dots (van Scheepen 1991).

'**Princess Alexandra**', a name applied by Gary Dunlop to a dwarf, vigorous and very late-flowering cultivar, possibly of Dutch origin, that had been wrongly identified in cultivation as 'Citronella'. It is about 2 feet (60 cm) tall and has rich yellow flowers about $1^3/_{16}$ inches (3 cm) across, with several small dark blotches around the center. Its appearance does match the description but does not match the watercolor illustration by E. A. Bowles, which shows a large, substantial plant similar to his illustration of 'Queen Alexandra'. Both illustrations are hung on the half-landing in the stairwell of Cory Lodge, Cambridge University Botanic Garden.

'**Princess Mary**', 1918, Henley. 'Lord Nelson' × 'Pageant' (parentage from J. E. Fitt notebooks). Very beautifully shaped flower with pale yellow tepals with red tips, with three of the tepals very bright crimson on the reverse side. Never before shown according to Fitt (1920). Highly commended by the RHS, 1939, after trial.

'**Profusion**', 1897, Lemoine. Stems are tall, upright and very well branched; the flowers are small, erect and brick orange with a brown center, and often semidouble.

'**Prolificans**', producing offshoots, date and breeder unknown. A vigorous plant with small, burnt orange, trumpet-shaped flowers.

'**Prometheus**' (Plate 15), 1904, Davison. ('George Davison' × 'Germania') × 'George Davison'. Much the largest-flowered cultivar of montbretia, each flower being about $3^{1}/_{2}$ inches (9 cm) across, of perfect shape and deep orange with a red ring around the center; about 4 feet (1.2 m) tall. RHS Award of Merit (in which 'George Davison' was miscalled 'Davisoni'), 1906; KAVB Award of Merit, 1908.

'**Pyramidalis**', pyramid-like, 1884, Lemoine. Erectly held, salmon-orange flowers.

'**Quantreau**', alluding to the French liqueur Cointreau, 2003, Dunlop. *Crocosmia* ×*crocosmiiflora* cultivar × *C. masoniorum*. Large, soft orange flowers of great substance with a pale yellow center, facing forward on a cascading stem about 3 feet (90 cm) long, which relatively short pleated foliage.

'**Queen Adelaide**', 1912, Henley. 'George Henley' × 'Prometheus'. Deep orange shaded with red on the outer side of the tepals; strong grower. Flowers 3–4 inches (7.5–10 cm) in diameter. RHS Award of Merit, 1913. KAVB Award of Merit, 1913.

'**Queen Alexandra**' (Plate 14), 1917, Henley. 'Queen Boadicea' × 'Pageant' (parentage from J. E. Fitt notebooks). Choice cultivar with erect habit, 3–4 feet (90–120 cm) high; large, reflexed, light golden orange flowers with distinct crimson bars. RHS Award of Merit, 1918.

'**Queen Boadicea**', 1919, Fitt. 'George Henley' × 'Pageant' (parentage from J. E. Fitt notebooks). Orange-copper, with very large flowers and beautifully rounded tepals; vigorous grower (Fitt 1920). Very vigorous, much-branched stems nearly 4 feet (1.2 m) high, carrying large and massive flowers, well expanded, of deep orange shading to copper (Wallace 1926).

'Queen Charlotte', 1918, Henley. 'George Henley' × 'Pageant' (parentage from J. E. Fitt notebooks). Large, handsome, dark orange flowers with pale centers. Good branching habit, 2–3 feet (60–90 cm) high (Fitt 1920).

'Queen Elizabeth', 1913, Henley. 'George Henley' × 'Pageant'. Brilliant shade of coppery red, paler in the center. Large flowers with broad, rounded tepals. RHS Award of Merit, 1915.

'Queen Mary', 1917, Henley. 'George Henley' × 'Tricolor' (parentage from J. E. Fitt notebooks). Fine cultivar growing about 3 feet (90 cm) tall, with a branching habit; beautiful, large, light orange flowers with crimson markings around the center, borne on stout dark stems. RHS Award of Merit, 1918.

'Queen Mary II', synonym of 'Columbus'. The name 'Queen Mary II' was applied by Gary Dunlop in 1995 to a Dutch cultivar of uncertain name that resembled the original 'Queen Mary'. It was presumably raised by Jac M. van Dijk but had arrived in cultivation in the United Kingdom under several different names. Willem Heemskerk, who bought the nursery stock from van Dijk, renamed it because he was unable to identify it.

'Queen of Spain', 1915, Henley. 'Queen Adelaide' × 'George Henley' (parentage from J. E. Fitt notebooks). The flowers of this very fine cultivar are $3^{1}/_{2}$ inches (9 cm) across and the color, somewhat deeper than that of 'Star of the East', is orange-red. The spike axis and bracts are dark red. The flowers are well expanded and borne above the foliage. One of the first to flower, usually in midsummer (July; Fitt 1924). RHS Award of Merit, 1916.

'Rachael', 1933, Fitt? Large red flowers (Prichard 1935).

'Rayon d'Or', golden ray, 1888, Lemoine. Large flowers are perfectly shaped, with broad tepals like a species *Gladiolus;* they are ochre-yellow with darkish brown spots at the base of the tube—like a *Tigridia* species, very special. A very dainty charming flower, not large (sic) but distinct and pretty. The buds are quite a warm reddish orange, but as the flower opens this gives way to almost uniform apricot yellow. The red remains in part but there is less of it;

the tepals are very firm, and in the center of the flower there is a ring of orange-red (Gumbleton 1905).

'**R. C. H. Jenkinson**', 1936, Fitt. Compact habit, with flower stems about 3 feet (90 cm) tall, freely produced; flowers 3^1/$_2$ inches (9 cm) wide, rich orange-scarlet, zoned at the throat, orange flushed scarlet. RHS Award of Merit, 1939, after trial, submitted in 1937.

'**R. C. Notcutt**', 1926, Fitt. 'His Majesty' × 'Tangerine'. Large-flowered, deep fiery orange, shading to clear yellow at the center. RHS Award of Merit, 1927.

'**Red Indian**', 1918, Henley. 'Stella' × 'Lord Nelson' (parentage from J. E. Fitt notebooks). Deep crimson flowers, which are the darkest of all the Earlham seedlings (Cowley 1918). Very fine dark red, center slightly shaded yellow (Fitt 1928).

'**Red King**', 1926, Fitt. Fine red flowers (Prichard 1929).

'**Red Knight**', 1928, Fitt. Large star-shaped flowers of a rich vermilion shading to gold at center, the center heavily overlaid with maroon, the reverse of the tepals dull crimson; a fine sturdy cultivar of vigorous constitution (Barr 1930, Cowley 1929).

'**Red Star**', 1992, Knuckley. An undistinguished small, red-flowered cultivar with a yellow eye; almost certainly of Dutch origin.

'**Rheingold**', Rhine gold, 1909, Pfitzer. Very large, bright golden yellow flowers with carmine blotches in the center, opening from orange buds. Listed as new by Pfitzer (1911). KAVB Award of Merit, 1938.

'**Rosamund**', pre-1927, Fitt? Rosy salmon flowers. Listed in the Daisy Hill Nursery catalogue in 1928.

'**Rosemary**', 1932, Fitt. 'Queen Charlotte' × 'R. C. Notcutt'. Flowers 3 inches (7.5 cm) across, brilliant orange-scarlet blotched crimson, tepals broad, throat golden yellow. RHS Award of Merit, 1933.

'**Rose Queen**', 1997, Dunlop. Flowers about 2 inches (5 cm) across, with yellow-centered orange flowers that quickly turn rose pink in sun.

'**Rose Seedling**', synonym of 'E. A. Bowles', grown under the former name by Helen Dillon and other good gardeners around Dublin, Ireland.

'**Rowallane Apricot**', 1990, breeder unknown. *Crocosmia masonio-*

rum. A soft apricot orange form with a much longer inflorescence than normal but very slow to increase. Named after the garden in which it originated, Rowallane Garden in Northern Ireland, now owned by the National Trust, where it occurred as a self-sown seedling resulting from a cross of 'Rowallane Orange' and 'Rowallane Yellow'.

'Rowallane Orange', 1995, breeder unknown. *Crocosmia masoniorum.* Considered superior because it is a slower-growing selection of the species.

'Rowallane Yellow', 1980, breeder unknown. *Crocosmia masoniorum.* A good chrome yellow form, presumed to be a naturally occurring sport but, unlike the yellow-flowered form of *Chasmanthe floribunda* called variety *duckittii,* does not reproduce true from seed. RHS Award of Garden Merit, 2002.

'Rowden Bronze', synonym of 'Coleton Fishacre'.

'Rowden Chrome', synonym of 'George Davison'.

'Royal Flush', 1997, Mahir. *Crocosmia masoniorum* × ? Apricot-colored flowers flushed heavily with carmine from the center and edges, well presented from contrasting purple stems 30 inches (75 cm) tall.

'Ruby Glow', pre-1932, Fitt. Large, open flowers 2½ inches (6 cm) across, rich crimson-rose flushed with orange and with a golden sheen; very beautiful (Wallace 1932).

'Rubygold', 1999, Dunlop. *Crocosmia paniculata* × *C. pottsii.* Forward-facing, rich red flowers with a golden suffusion inside borne on stems more than about 39 inches (1 m) tall, above the pleated foliage.

'Ruby King', 1926, Fitt. Large, open, star-like flowers measuring 2½ inches (6 cm) across, of a rich crimson-rose with orange flush and crimson sheen; a refined, beautiful flower. Height 30 inches (75 cm) (Barr 1932).

'Ruby Velvet', 1997, Dunlop. *Crocosmia masoniorum* × *C. paniculata.* Flowers deep, rich shade of red that is difficult to focus the eye on and almost impossible to photograph; plants very slow to increase.

'R. W. Wallace', 1926, Fitt. Charming, the flowers a clear golden yellow, reverse deep orange, center white and gold shaded; erect habit (Fitt 1928).

'Saffron Queen', 1996, Dunlop. Large, soft, apricot orange flowers more than 2 inches (5 cm) across, with pale yellow eyes and un-usual, very broadly flaring perianth tubes, on stems about 39 inches (1 m) tall, with curving side branches. It flowers earlier than most large-flowered *Crocosmia* ×*crocosmiiflora* cultivars; ro-bust and very vigorous.

'St Botolph', 1906, Davison. Growing 4 feet (1.2 m) high and with much-branched, very strong stems held quite erect, producing quantities of flowers, each 3 inches (7.5 cm) across and opening quite flat; with the largest flowers of all the yellow cultivars at the time.

'St Clements', 2002, Dunlop. *Crocosmia pottsii* × 'Sulphurea'. Rela-tively small-flowered, with upward-facing trumpet-shaped flow-ers that are lemon yellow inside and middle orange outside, fading red, on widely branched stems about 3 feet (90 cm) tall, thus pro-viding a bright color combination in an open, airy display.

'Saracen', 1984, Browne. 'Jackanapes' × 'Solfatare'. Bright red flowers about 1¹/₂ inches (38 mm) across, with small yellow centers; flow-ering about 30 inches (75 cm) high, just clear of the bronze foli-age.

'Scarlatti', 2000, Dunlop. *Crocosmia paniculata* × *C.* ×*crocosmiiflora*. Robust, with bright red, trumpet-shaped flowers that flare open and flower at about 30 inches (75 cm) high.

'Sensation', 1923, Fitt? One of the most distinctive, rich orange, shaded crimson (Prichard 1925).

'Severn Sunrise', 1990, Cattermole and Durrant. *Crocosmia maso-niorum* × *C. paniculata*. Vigorous, tightly clumping. The bright orange flowers have yellow eyes, are well presented on stems about 3 feet (1 m) tall. In sunlight the flowers quickly fade a rich middle pink. RHS Award of Garden Merit, 2002.

'Shocking', 1997, Dunlop. *Crocosmia masoniorum* × *C. paniculata*. Flowers open bright orange with a yellow eye, quickly turning a vivid shade of shocking pink in full sun.

'Sir Matthew Wilson' (Plate 13), 1927, Fitt. 'Princess' × 'Queen Charlotte'. Very brilliant orange-red that in good light looks almost scarlet, with a small golden yellow zone around the center with carmine markings. RHS Award of Merit, 1928.

'**Soleil Couchant**', sunset, 1889, Lemoine. Very floriferous dwarf plant, with erect stems and brilliant golden yellow flowers of great size, presented full face. RHS Award of Merit, 1895.

'**Solfatare**' (Plate 14), 1886, Lemoine. Large flowers of Naples yellow, displaying themselves full face on almost erect stems 26 inches (65 cm) tall; a very distinctive and attractive cultivar. 'Solfaterre' is a long-running misspelling; Nelson (1999) identified its first use in *The Gardeners' Chronicle* only a year after 'Solfatare' was first released and thus established unambiguously the correct spelling. This did not, however, explain why the misspelled name persisted for well over a century, with both spellings used concurrently by different nurseries in England in the 19th century and, more erratically, in the 20th. The answer almost certainly lies in the fact that the Lemoine catalogue was printed in both French and English editions. At least one early English edition survives with the incorrect spelling, which is not a translation. It is likely that this spelling error occurred in the very first listing of 'Solfatare' in the English catalogue and was never corrected. Other variants include 'Solfataire', 'Solfatara' and 'Solfaterra'. *Solfatara* is both the English and Italian translation of the French *solfatare; solfatara* is a geological term for fumarole, a gaseous volcanic vent that discharges sulphurous fumes. The name derives from the volcanic region just north of Naples, which was a well-known tourist attraction on the Grand Tour in Victorian times. Lemoine made no mention of the plant's having distinctively colored foliage. The choice of name for a plant with sulphur yellow flowers and smoky green foliage no doubt alluded to this geological feature. RHS Award of Garden Merit, 1993.

'**Solfatare Coleton Fishacre**', see 'Coleton Fishacre'.

'**Sonate**', sonata, 1993, van Dijk. Listed as new by van Dijk (1995). Flowers yellow-brown. Plants sold under this name are a vigorous hybrid, *Crocosmia masoniorum* × *C. pottsii,* with orange flowers with a broad brown ring around the eye. See also 'Marcotijn'.

'Spitfire', 1970, Bloom. *Crocosmia masoniorum* × *C.* ×*crocosmiiflora*. Plants about 30 inches (75 cm) high, with widely opening, vivid, fiery orange flowers.

'Sproston Gem', 1935, Hammond. No description traced.

'Sproston Glory', 1935, Hammond. Broad tepals, dark orange-red outside, orange-brown inside, fading to a pinkish tinge with a wide ring of red-brown around the wide yellow throat (Thomas 1987).

'Starbright', 2003, Dunlop. An openly branched, robust plant with white-centered, bright yellow flowers about 1½ inches (38 mm) in diameter; plant about 30 inches (75 cm) tall.

'Starfire', 1985, Browne. 'Jackanapes' × 'Solfatare'. Red flowers about 1¼ inches (32 mm) across, with pale yellow centers; about 2 feet (60 cm) tall, with bronze foliage.

'Star of the East' (Plate 13), 1910, Davison. A splendid acquisition and doubtless the finest montbretia raised at the time. The flowers open a full 4 inches (10 cm) across and are a lovely golden orange. The buds are prettily tinged with scarlet. The center of the flower is very pale yellow (Arnold 1933). RHS First Class Certificate, 1912; Award of Garden Merit, 2002. KAVB Award of Merit, 1913.

'Strawberry Cream', 2002, Dunlop. *Crocosmia paniculata* × *C.* ×*crocosmiiflora*. Flowers bright red, medium-sized, forward facing, with a relatively large but irregular cream center. Stems about 3 feet (90 cm) tall and leaves pleated.

'Stuttgardia', 1902, Pfitzer. Dwarf; very large, clear yellow flowers. Listed as new (in lowercase) by Daisy Hill Nursery (1906) but available only by the dozen.

'Sulphurea', sulphur yellow, 1884, Lemoine. Taller than *Crocosmia pottsii* and with a greater number of spikes. The flowers are also larger, tubular and a dark chrome yellow.

'Sultan', 1984, Browne. 'Jackanapes × 'Solfatare'. Rich red flowers about 1½ inches (38 mm) across, with purple-spotted yellow centers; about 30 inches (75 cm) tall with bronze foliage.

'Sunset', 1997, Mahir. Glaucous stems 30 inches (75 cm) tall bear rounded flowers more than 1½ inches (38 mm) across and of an unusual ochre-apricot color flushed brown at midsegment and with a heavy purple dash at the base of the lower tepals. The bracts are pale violet.

'**Sunshine**', 1918, Henley. 'George Henley' × 'Tricolor' (parentage from J. E. Fitt notebooks). Flowers of a distinctive shape, cherry scarlet merging to gold. Good growth habit; 30 inches (75 cm) high (Fitt 1920). The flowers are cherry scarlet and gold; named as such because it looked so brilliant in morning sun (Cowley 1918).

'**Superba**', date and breeder unknown. *Crocosmia pottsii*. Red and yellow is the full description (Jackman 1912). Also called (?) '*C. pottsii* Improved' and possibly a synonym for 'Grandiflora'.

'**Surprise**', pre-1906, Welker. Dwarf plants with old-gold-colored flowers flushed with carmine on the inner tips and the outside, with red stripes at the base of the tube (Grignan 1906).

'**Sybil**', pre-1963, van Nieuwkoop. Light apricot yellow flowers with center burnt orange (van Scheepen 1963).

'**Talisman**', 1888, Lemoine. Flowers are completely erect but only partially open; they are blood vermilion in color, with scattered purple spots in the throat.

'**Tangerine**', 1918, Henley. 'Queen Elizabeth' × 'George Henley' (parentage from J. E. Fitt notebooks). Broad, pointed tepals, rich orange shaded to yellow in the center (Fitt 1920).

'**Tangerine Queen**', 1993, Dunlop. *Crocosmia masoniorum* × *C. pottsii*. Relatively vigorous, with upright stems and compact, well-presented, rich orange inflorescences; about 3 feet (90 cm) tall, with neat pleated foliage; of Dutch origin.

'**Ternace Tomato**', 1994, Fenwick. Red-flowered, evidently a garden escape, about 2 feet (60 cm) tall.

'**Tête Couronne**', crowned head, 1898, Lemoine. Green foliage turning bronze; very large yellow-orange flowers with a large maroon ring ³/₈ inch (1 cm) broad around the yellow throat.

'**The King**', 1912, Hayward. Brilliant orange-red flowers, 3 inches (7.5 cm) across, opening flat on branching stems; tall and robust (Hayward 1914).

'**The Prince**', 1919, Fitt. Similar in coloring and size to 'His Majesty' but with a circle of dark red spots around the base of each segment (Fitt 1920).

'Thor', 1912, Hayward. Brilliant, fiery red, trumpet-shaped flowers on stems 3–4 feet (90–120 cm) tall (Hayward 1914).

'Tiger', 1993, Lewis. *Crocosmia pottsii* × *C. masoniorum*. Rich orange flowers with a distinct dark stripe in the middle of each tepal (Lewis 1994).

'Tigridie', *Tigridia*-like, 1889, Lemoine. Flowers are presented full face on erect stems; they are a rich orange, with a broad ring of dark brown spots in the throat.

'Toison d'Or', golden fleece, 1900, Lemoine. Large, fully open, erect, golden yellow flowers on straight stiff stems.

'Tragédie', tragedy, 1900, Lemoine. Branched spikes of enormous round, open, dark orange-yellow flowers with a broad ring of black velvet occupying the central half of the flower, and with somber green foliage—a very good name. It is an intensely dark hybrid, probably the darkest then in cultivation. The flower is deep orange, the tepals broad and a warm maroon-purple in the lower halves. The buds are quite a lurid color, almost black, but dull orange toward the apex. It is a very handsome flower and, associated with certain plants, would be very effective (Gumbleton 1905).

'Transcendant', transcendent, 1888, Lemoine. Flowers are large with very large tepals, orange-vermilion on the outside, bright vermilion on the inside and with a yellow throat. It is the most floriferous of all crocosmias and flowers during the whole summer and autumn; it is most useful in pots.

'Triomphe de Paris', pre-1895, Welker. No description traced (Journal de la Société Nationale d'Horticulture de France 1895: 544).

'Tropicana', 2003, Dunlop. *Crocosmia masoniorum*. Large-flowered form with rich orange flowers, with the uppermost tepal of each flower strongly gilded.

'Turban', 1895, Lemoine. Large, well-presented flowers are ochre-yellow on the outside and dark yellow inside; a very broad ring of blood purple markings, which show though to the outside, surrounds the center, with a yellow circle in the throat.

'Una', 1918, Henley. A charming cultivar suggestive of the form of

'Citronella'; of a pale orange shade with crimson reverse, the inner surface of the flower characterized by crimson markings at the base of the tepals (Fitt 1920). RHS Award of Merit, 1919.

'Venus', pre-1991, van Dijk. Outer tepals and perianth tube mandarin red, inner tepals dark buttercup yellow, inside Indian yellow, tipped nasturtium red (van Scheepen 1991).

'Vermeer', 1918, Jan Roes. Clear orange flowers with a brown ring around the eye. Listed without description in 1920 but with description in 1923.

'Versicolor', multicolored, pre-1906, Welker. Very large flowers with pale yellow centers, flushed carmine-orange at the tips of the tepals (Grignan 1906).

'Vesuv', Vesuvius, 1905, Pfitzer. Bright scarlet-red with dark yellow throat (Pfitzer 1911).

'Vesuve', Vesuvius, 1894, Krelage. Very dark, fiery red (Krelage 1896).

'Victoria', 1988, Ursem. Large Indian orange flowers edged tomato red with a lemon yellow throat (van Scheepen 1991).

'Victor Welker', pre-1906, Welker. Attractive plants with internally vivid and externally dark orange flowers; there is a dark brown disk in the center, with a paler halo (Grignan 1906).

'Volcan', volcano, 1897, Lemoine. Stems are more than 39 inches (1 m) tall. Flowers are large and perfect, with broad tepals, dark cadmium yellow with a yellow center with light purple spots. The plant will eventually form tall clumps.

'Volcano', 1894, breeder unknown. Flowers dark red, yellow in the center, with purple blotches (Krelage 1896).

'Voyager', 1993, van Dijk. Rich chrome yellow self-sown seedling flowers open flat, about 2½ inches (6 cm) across; plants robust, relatively slow to increase and flowering at about 30 inches (75 cm) high.

'Vulcan' (Plate 13), 1897, Leichtlin? *Crocosmia paniculata* × *C. aurea*. Very tall, similar to *C.* ×*crocosmoides;* flowers deep orange-red (Daisy Hill Nursery 1906). 'Vulcan' has radially symmetric flowers and a yellow eye. At 4 feet (1.2 m) it is the tallest of the five named cultivars of *C.* ×*crocosmoides* still in cultivation.

'**Vulcan**', 1970, Bloom. *Crocosmia masoniorum* × *C. paniculata*. Very similar to 'Bressingham Blaze' but the flowers are a shade darker, as is the slightly broader foliage.

'**Walberton Red**', pre-1980, Tristram. A hybrid of unknown parentage with *Crocosmia masoniorum* dominant. A slowly increasing red-flowered plant with sickly looking, dusky foliage. The original stock was destroyed by the breeder as he considered the plant to be inferior, but not before some material had been distributed. He suggested the name 'Old Hat' should be used for the surviving plants, but the original name, having been published (Thomas 1982), must stand.

'**Walberton Stripey**', pre-1980, Tristram. With striped red and yellow flowers, as the name implies, submitted to the RHS trial of *Crocosmia* in the late 1980s; subsequently destroyed by the breeder.

'**Walberton Yellow**', pre-1980, Tristram. A hybrid of unknown parentage with *Crocosmia masoniorum* dominant. Slow-growing, with the appearance of a small, chrome yellow form of *C. masoniorum*. Protected by Plant Breeder's Rights.

'**Wasdale Red**', 1994, Fenwick. Red-flowered, evidently a garden escape, about 30 inches (75 cm) tall.

'**Westwick**', 1905, Davison. Very distinctive orange-red with a clear yellow eye surrounded with a dark circle of maroon.

'**Zeal Giant**', 1990, Jones. *Crocosmia paniculata* × *C. masoniorum*. Rich, pale orange-flowered; almost 6 feet (1.8 m) tall but slow to increase and with enormous corms.

'**Zeal Tan**', 1991, Jones. Attractive, with medium-sized, bright orange-red flowers with a fine yellow eye; just over 2 feet (60 cm) tall.

'**Zeal Unnamed**', 1992, Jones. *Crocosmia paniculata* × *C. masoniorum*. Vigorous, deeply orange-flowered, about 56 inches (1.4 m) tall. The flower panicles are presented at an angle slightly below the horizontal.

Specialist Nurseries for *Crocosmia*

There are a growing number of plant finders available that provide the most effective way of finding out where to obtain specific plants. The original one, covering the British Isles, devised by the late Chris Philip, is now published by the Royal Horticultural Society; information in the *RHS Plant Finder* is available on the Web. The New Zealand PlantFinder is also online, with a subscription charge. There are printed plant finders for Australia (*Aussie Plant Finder*) and the Netherlands (*Plantenvinder voor de Lage Landen*). The largest plant finder is *PPP-Index,* which covers much of Europe. In the United States, there is the Andersen Horticultural Library's *Source List of Plants and Seeds* and, for western North America, *The Plant Locator,*® *Western Region.*

Australia

Lambley Nursery
'Burnside'
Lester Road, Ascot
Victoria 3364

Tesselaar Bulbs and Flowers
357 Monbuck Road
Silvan
Victoria 3795

Canada

Phoenix Perennials
3380 Sixth Road
Richmond
British Columbia V6V 1P3

France

Bulbes d'Opal
Cidex 528
384 Boerenweg Ouest
59285 Buysscheure

France (continued)

Bulb'Argence
Mas d'Argence
30300 Fourques

Establissement Simon Arbuthnot
Bourdelas
24800 Saint-Jory-de-Chalais

Germany

Staudengärtnerei Re-natur
Plöner Strasse 10
24619 Bornhöved

Ireland

Brookwood Nurseries
18 Tonlegee Road,
Coolock
Dublin 5

The Netherlands

Kwekerij Davelaar
Bloembollen & Bolbloemen
Davelaar 4
3931 PS Woudenberg

Kwekerij De Hessenhof
Hessenweg 41
6718 TC Ede

Coen Jansen
Ankummer Es 13a
7722 RD Delfsen

United Kingdom

Avon Bulbs
Burnt House Farm
Mid-Lambrook
South Petherton
Somerset TA13 5HE

Ballyrogan Nurseries
The Grange
Ballyrogan
Newtownards
Co. Down BT23 4SD

Blooms of Bressingham
Bressingham
Diss
Norfolk IP22 2AB

Bridgemere Nurseries
Bridgemere
Nantwich
Cheshire CW5 7QB

Broadleigh Bulbs
Bishops Hull
Taunton
Somerset TA4 1AE

Cally Gardens
Gatehouse of Fleet
Castle Douglas
Scotland DG7 2DJ

Cotswold Garden Flowers
Sands Lane
Badsey
Evesham
Worcestershire WR11 5EZ

Fir Tree Farm Nursery
Tresahor
Constantine
Falmouth
Cornwall TR11 5PL

Four Seasons
Forncett St Mary
Norwich
Norfolk NR16 1JT

Hillview Hardy Plants
Worfield
Bridgnorth
Shropshire WV15 5NT

Rowden Gardens
Brentor
Tavistock
Devon PL19 0NG

United States of America

Digging Dog Nursery
P.O. Box 471
Albion
California 95410

Heronswood Nursery
7530 N.E. 288th Street
Kingston
Washington 98346

Joy Creek Nursery
203000 N.W. Watson Road
Scappoose
Oregon 97056

Odyssey Bulbs
8984 Meadow Lane
Berrien Springs
Michigan 49103

Glossary

abaxial Of the side or face of an organ that faces away from the axis; usually the lower face

actinomorphic Radially symmetric

anther Pollen-bearing organ, top of part of a stamen (male part of the flower), borne at the top of a stalk, the filament

apiculum (adjective, **apiculate**) A short point

bi- Prefix denoting two

bract (adjective, **bracteate**) Leaf-like structure lacking an axillary bud (a bud in the angle between a leaf or bract and the stem bearing it) and often subtending a flower or inflorescence, usually differing from foliage leaves in form or size

capsule Dry fruit formed from two or more united carpels (that is, with two or more locules) and dehiscent at maturity to release the seeds

cataphyll Scale-like leaf, usually associated with a vegetative propagating organ like a corm or perennating bud

cauline Borne along the stem or trunk

chalaza (adjective, **chalazal**) End of the seed opposite the micropyle (the opening at the base of the ovule though which the sperm cells enter the embryo sac) and the point of entry of the vascular tissue into the ovule

clade An evolutionary lineage in which all the members have a common ancestor

clone Single genetic line (genotype) maintained through vegetative reproduction

corm Underground bulb-like storage organ derived from stem tissue

cormlet Small corm

decurrent Extending downward below the point of insertion

dehiscent Breaking open at maturity to release the contents

depressed Flattened as if pressed down from the top or end

distichous Arranged in two opposite rows

divaricate (adjective, **divaricately**) Widely spreading, often almost horizontally

exine Outer layer of the wall of a pollen grain

exserted Protruding

-fid Suffix denoting divided, usually for about one-third of the length

filament Stalk of a stamen

filiform Thread-like

flexuose Bent from side to side in a zigzag form

glaucous Bluish green with whitish bloom

globose Nearly spherical

herbarium Museum of dried plant specimens, pressed flat and mounted on sheets of paper or light cardboard

holotype The single herbarium specimen or illustration upon which the name of a taxon (a naturally occurring taxonomic group of plants) is based, usually cited as such when a plant is first described

inflexed Bent abruptly upward or forward

inflorescence Group or arrangement in which flowers are borne on the plant

isotype A duplicate of a holotype

karyotype The appearance of the chromosome complement of an organism

lanceolate Broadest in the lower half and tapering toward the tip, about four times as long as broad

lectotype Deliberately chosen type (herbarium specimen or illustration) of a taxon when the holotype is uncertain and was not designated when the plant was first described

locule Enclosed compartment within an organ, usually of the ovary

midrib Central vein (the midvein) of a leaf

nectary Multicellular gland secreting nectar

neotype Designated new type (herbarium specimen or illustration) of a taxon when the original type is lost

oblique Off-center or not symmetric

oblong Flat, with parallel sides and symmetrically rounded ends

obtuse At an angle of more than 90°

ovoid Egg-shaped in three dimensions

ovule Organ within the ovary that houses the embryo sac

panicle (adjective, **paniculate**) Branched or compound inflorescence

papilla (adjective, **papillate**) Small, elongate projection on the surface of an organ

perforate Of a surface with small perforations; specifically in pollen, with minute pores in the outer layer

perianth Calyx (sepals or outer sterile whorl) and corolla (petals or inner sterile whorl) of a flower, particularly when the two are similar

petiole Stalk of a leaf

protologue Original formal description and naming of a species, sub-species or group of any other taxonomic rank

raphe (adjective, **raphal**) Dorsal ridge of a seed, which carries the vascular trace from the base of the ovule (the micropylar end) to the base of the embryo at the opposite end of the seed

recurved Curved or curled downward or backward

reflexed Bent abruptly downward or backward

rugose Deeply wrinkled

scabrate Minutely roughened by microscopic hairs

secund With all the parts grouped on one side or turned to one side

semi- Prefix denoting half

septum (adjective, **septal**) Partition, particularly between the locules of the ovary

serrate Toothed, with asymmetric teeth pointing forward

sessile Without a stalk

spathulate Spatula-shaped, thus broadest shortly below the rounded tip

spike (adjective, **spicate**) Unbranched, indeterminate inflorescence in which the flowers are sessile or without stalks on the axis,

stamen One to many in number, comprising the male portion of the flower, usually consisting of a filament and anther; in the **Iris** family there are nearly always three stamens per flower

stigma (adjective, **stigmatic**) Receptive tissue to which the pollen is delivered, usually at the tip of the style, or the style branches

stolon Prostrate or trailing stem, usually underground, that produces roots or corms at the nodes

stoloniferous Bearing a stolon

style Tissue connecting the ovary to the stigma, sometimes branched, as in *Chasmanthe* and *Crocosmia*

sub- Prefix denoting approximately, more or less

subepidermal Under the epidermis or outer covering

subpatent Halfway to being fully spreading

subversatile Nearly versatile, in reference to an anther that rotates to a limited extent at its point of insertion at the top of the filament

sulcus Aperture of a pollen grain, an area usually lacking exine and elliptic in shape

syntype One of two or more herbarium specimens or other elements cited when a species is first described and when none was specifically designated the holotype

tepal Member of the nonfertile whorls of floral parts when both whorls, inner and outer, are similar rather than consisting of a green, leafy outer whorl of sepals (the calyx) and an inner whorl of colored petals (the corolla)

tri- Prefix denoting three

type See holotype, isotype, lectotype, neotype, syntype

unifacial Of leaves that have both faces anatomically identical, typically flattened parallel to the stem axis

zygomorphic Of a flower, bilaterally symmetric

Bibliography

Albyn. 1939. Montbretias—recent hybrids. Gardening Illustrated 61: 623.

ALW. 1931. Montbretias. The Gardeners' Chronicle 89: 163.

Archibald, J. 2000. 'Raiser Unknown': Eric Smith—a plantsman. The Hardy Plant 22: 94–108.

Arnold, R. E. 1933. *Montbretia* Star of the East. Gardening Illustrated 55: 500.

Arnott, S. 1899a. Montbretias and tritonias in Scotland. The Gardeners' Chronicle 25: 412–413.

Arnott, S. 1899b. *Antholyza paniculata*. The Garden 56: 192.

Arnott, S. 1900. *Antholyza paniculata*. The Garden 58: 200.

Arnott, S. 1904. *Antholyza paniculata* 'Major'. The Garden 66: 348.

Arnott, S. 1909. *Antholyza paniculata*. The Garden 73: 442–443.

Arnott, S. 1910a. Hardy flower chat [notes on *Montbretia* 'Comet' and *Antholyza paniculata*]. The Gardeners' Magazine 53: 805.

Arnott, S. 1910b. An interesting autumn flower. The Garden 75: 597.

Backhouse, J. 1911. [Bulb catalogue, page 32] James Backhouse and Son, York.

Bailey, L. H. 1914. The Standard Cyclopedia of Horticulture, vol. 1. Macmillan, New York.

Baker, H. G., and I. Baker. 1983. Floral nectar sugar constituents in relation to pollinator type, pp. 117–141 *in* C. E. Jones and R. J. Little (editors), Handbook of Experimental Pollination Biology. Scientific and Academic Editions, New York.

Baker, J. G. 1877a. *Montbretia pottsii*. The Gardeners' Chronicle 8: 424.

Baker, J. G. 1877b. Systema Iridacearum. Journal of the Linnean Society, Botany 16: 61–180.

Baker, J. G. 1883. *Tritonia pottsii.* Curtis's Botanical Magazine 109: plate 6722.

Baker, J. G. 1888. *Crocosmia aurea* var. *maculata.* The Gardeners' Chronicle 4: 407, 565 and figure 80.

Baker, J. G. 1892. Handbook of the Irideae. Reeve and Company, Ashford, Kent.

Baker, J. G. 1896. Iridaceae. Flora Capensis 6: 7–171.

Barr, P. 1924–1936. [General bulb catalogues] Barr and Sons, London.

Beissner, L. 1892. Der Leichtlinsche Garten in Baden-Baden. Gartenflora 41: 607–608.

Bloom, A. 1965. Hardy Plants of Distinction. Collingridge, London.

Bloom, A. 1991. Hardy Perennials. Batsford, London.

Bolus, H. M. L. 1921. Novitates Africanae. Annals of the Bolus Herbarium 3: 70–85.

Bolus, H. M. L. 1926. Novitates Africanae. Annals of the Bolus Herbarium 4: 37–55.

Bolus, H. M. L. 1933. Plants—new and noteworthy. South African Gardening 23: 46–47.

Bowles, E. A. 1917. Recollections of the garden at Earlham Hall. The Garden 90: 114–115.

Brown, N. E. 1932. Contributions to a knowledge of the Transvaal Iridaceae. 2. Transactions of the Royal Society of South Africa 20: 261–280.

Bull, W. 1891. [Catalogue] Chelsea, London.

Burbidge, F. W. T. 1890a. *Montbretia pottsii.* The Gardeners' Chronicle 7: 301.

Burbidge, F. W. T. 1890b. Montbretias. Journal of Horticulture, Cottage Gardener and Home Farmer, ser. 3, 21: 226.

Cowan, H. (editor). 1917. September flowers at Earlham Hall. The Garden 90: 372–373.

Cowley, H. (editor). 1918. The Earlham montbretias. The Garden 91: 339.

Cowley, H. (editor). 1929. Norfolk and montbretias. Gardening Illustrated 51: 621.

Cornut, J.-P. 1635. Canadensium Plantarum Aliarúmque Nondum Editarum Historia. Simon le Moyne, Paris.

Daisy Hill Nursery. 1898–1970. [Catalogues] Newry, Northern Ireland.

Davison, G. D. 1905. The new montbretias. The Garden 68: 169–170.

Desmond, R. 1994. Dictionary of British and Irish Botanists and Horticulturists, ed. 2. Taylor and Francis, and Natural History Museum, London.

De Vos, M. P. 1982a. The African genus *Tritonia* Ker-Gawler (Iridaceae): part 1. Journal of South African Botany 48: 105–163.

De Vos, M. P. 1982b. Die bou and ontwikkeling van die unifasiale blaar van *Tritonia* en verwante genera. Journal of South African Botany 48: 23–37.

De Vos, M. P. 1983. The African genus *Tritonia* Ker-Gawler (Iridaceae): part 2. Sections *Subcallosae* and *Montbretia*. Journal of South African Botany 49: 347–422.

De Vos, M. P. 1984. The African genus *Crocosmia* Planchon. Journal of South African Botany 50: 463–502.

De Vos, M. P. 1985. Revision of the South African genus *Chasmanthe* (Iridaceae). South African Journal of Botany 51: 253–261.

De Vos, M. P. 1999a. *Chasmanthe.* Flora of Southern Africa 7(2: 1): 142–147.

De Vos, M. P. 1999b. *Crocosmia.* Flora of Southern Africa 7(2: 1): 129–138.

Dix, J. F. C. 1957. Bulb Growing for Everyone. Blandford, London.

Dunlop, G. 1999. Bright sparks. The Garden 124(8): 599–605.

Eliovson, S. 1955. South African Flowers for the Garden. Howard Timmins, Cape Town.

Fish, M. 1965. A Flower for Every Day. Studio Vista, London (reprinted 1973 by David and Charles, Newton Abbot).

Fish, M. 1966. Carefree Gardening. Collingridge, Southampton.

Fitt, J. E. 1920, 1924, 1928. Giant Earlham montbretias [catalogue]. Norfolk, England.

Fitt, J. E. 1932. Montbretias. The New Flora and Silva 4(3): 178–179.

Gasparrini, G. 1832. Osservazioni intorno ad alcune piante coltivate nel Real Orto Botanico di Boccadifalco, presso Palermo. Annali Civili del Regno delle Due Sicilie 1(4): 116–125.

GBM. 1904. *Montbretia* Chloris. The Garden 66: 203.

GBM. 1905. *Montbretia* Aurantiaca. The Garden 68: 119.

Gerard, J. N. 1897. *Antholyza crocosmoides.* Garden and Forest 10: 315–316.

Gerbeaux, F.-V. 1907, 1910. [Catalogues] Gerbeaux et Crouse, Nancy.

Germishuizen, G., and A. Fabian. 1977. Wildflowers of Northern South Africa. Fernwood Press, Cape Town.

Goldblatt, P. 1971. Cytological and morphological studies in the southern African Iridaceae. Journal of South African Botany 37: 317–460.

Goldblatt, P. 1993. Iridaceae. Flora Zambesiaca 12(4): 1–106.

Goldblatt, P. 1996. *Gladiolus* in Tropical Africa. Timber Press, Portland, Oregon.

Goldblatt, P. 2002. *Chasmanthe.* Flora of North America 26: 403–404.

Goldblatt, P., and J. C. Manning. 1990a. The Madagascan *Geissorhiza ambongensis* transferred to *Crocosmia* (Iridaceae–Ixioideae). Bulletin du Muséum National d'Histoire Naturelle, série 4, section B, Adansonia 12: 59–64.

Goldblatt, P., and J. C. Manning. 1990b. *Devia xeromorpha,* a new genus and species of Iridaceae–Ixioideae from the Cape province, S. Africa. Annals of the Missouri Botanical Garden 77: 359–364.

Goldblatt, P., and J. C. Manning. 1995. Phylogeny of the African genera *Anomatheca* and *Freesia* (Iridaceae–Ixioideae), and a new genus *Xenoscapa.* Systematic Botany 20: 161–178.

Goldblatt, P., and J. C. Manning. 1998. *Gladiolus* in Southern Africa. Fernwood Press, Cape Town.

Goldblatt, P., and J. C. Manning. 2000. Iridaceae, *in* O. A. Leistner (editor), Seed Plants of Southern Africa: Families and Genera. Strelitzia 10: 623–638. National Botanical Institute, Pretoria.

Goldblatt, P., A. Bari and J. C. Manning. 1991. Sulcus variability in the pollen grains of Iridaceae subfamily Ixioideae. Annals of the Missouri Botanical Garden 78: 950–961.

Goldblatt, P., J. C. Manning and P. Bernhardt. 1999. Evidence of bird pollination in the Iridaceae of southern Africa. Adansonia, série 3, 21: 25–40.

Grignan, G. T. 1906. Montbretias nouveaux. Revue Horticole 1906: 404–405.

Grignan, G. T. 1907. Montbretias hybrides. Revue Horticole 1907: 208–209.

Gumbleton, W. E. 1905. New montbretias. The Garden 68: 217.

Gunn, M., and L. E. Codd. 1981. Botanical Exploration of Southern Africa. A. A. Balkema, Cape Town.

Hayward, P. S. 1910. Montbretias. The Gardeners' Magazine 53: 320.

Hayward, P. S. 1914. The best montbretias. The Garden 138: 435–436.

Herbert, W. 1838. *Tritonia fucata*. Edwards's Botanical Register 24: plate 35.

Hinkley, D. 2003. Heronswood Nursery [catalogue], page 151. Kingston, Washington.

Hooker, J. D. 1847. *Tritonia aurea*. Curtis's Botanical Magazine 73: plate 4335.

Jackman, G. 1912. [Catalogue] G. Jackman and Sons, Surrey, England.

Ker-Gawler, J. B. 1802. *Antholyza aethiopica*. Curtis's Botanical Magazine 16: plate 561.

Ker-Gawler, J. B. 1809. *Antholyza aethiopica* var. ? Curtis's Botanical Magazine 29: plate 1172.

Klatt, F. W. 1867. Diagnoses Iridearum novarum. Linnaea 35: 377–384.

Klatt, F. W. 1882. Ergänzungen und Berichtigungen zu Baker's *Systema Iridacearum*. Abhandlungen der Naturforschenden Gesellschaft zu Halle 15: 335–404.

Klatt, F. W. 1894. Iridaceae Africanae. Conspectus Florae Africae 5: 143–250.

Klotzsch, J. F. 1851. *Tritonia* oder *Babiana aurea*. Allgemeine Gartenzeitung 1851: 293.

Kostelijk, P. J. 1984. *Crocosmia* in gardens. The Plantsman 5: 246–253.

Krelage, E. H. & Zoon. 1896– . [Catalogues] Haarlem.

Leichtlin, M. 1900. Third List of Roots and Plants. Baden-Baden.

Lemoine, E. 1900. *Crocosmia aurea* and tritonias (syn. montbretias). Journal of the Royal Horticultural Society 25: 128–132.

Lemoine, V. 1882–1910. [Catalogues] Lemoine, Nancy.

Lewis, A. 1994. The Abbey Nursery Catalogue. Devon, England.

Lewis, G. J. 1954. Some aspects of the morphology, phylogeny and taxonomy of the South African Iridaceae. Annals of the South African Museum 40: 15–113.

Lindley, J. 1828. *Antholyza aethiopica* L. var. *minor*. The Botanical Register 14: plate 1159.

Linnaeus, C. 1759. Systema Naturae, ed. 10. 2: 863.

Loudon, Mrs (Jane), 1841. The Ladies' Flower-Garden of Ornamental Bulbous Plants. London, W. Smith.

Macself, A. J. (editor). about 1930. Sanders' Encyclopaedia of Gardening, ed. 22. W. H. and L. Collingridge, London.

Manning, J. C., P. Goldblatt and D. Snijman. 2002. The Color Encyclopedia of Cape Bulbs. Timber Press, Portland, Oregon.

Marloth, R. 1917. Flora of South Africa, vol. 1. Darter Brothers, Cape Town.

Mason, M. 1913. Some flowers of eastern and central Africa. Journal of the Royal Horticultural Society 39(1): 8–16.

Masquelier, O. 1995. Victor Lemoine 1823–1911. Hortus 34: 77–89.

Maxwell, H. 1936, *Antholyza paniculata*. Gardening Illustrated 58: 616.

Milne-Redhead, E. 1948. *Crocosmia pauciflora*. Kew Bulletin 1948: 469.

Molyneux, E. 1907. Hybrid montbretias. The Garden 69: 28–29.

Morison, R. 1680. Plantarum Historiae Universalis Oxoniensis, Pars Secunda. Oxford, Teatro Sheldoniano.

Morren, C. J. E. 1852. *Antholyza bicolor*. La Belgique Horticole 2: plate 25/1.

Morren, C. J. E. 1881. Notice sur le *Montbretia crocosmiaeflora* (hybrida) de M. V. Lemoine ? *Montbretia aureo–pottsii*. La Belgique Horticole 31: 299–303, plate 14.

Nelson, E. C. 1993. Who was the author of *Montbretia crocosmiiflora*? Watsonia 19: 265–267.

Nelson, E. C. 1999. Lemoine's bronze-leaved montbretia: *Crocosmia* ×*crocosmiiflora* 'Solfatare'. The New Plantsman 6(2): 75–77.

Nicholson, G. (editor). 1888. The Illustrated Dictionary of Gardening, vol. 4. L. Upcott Gill, London.

Obermeyer, A. A. 1981. A new species of *Crocosmia*. Bothalia 13(3–4): 450–451.

O'Brien, J. 1890. *Crocosmia aurea* var. *maculata*. The Garden 38: 333.

Pax, F. 1893. *Tritonia cinnabarina*. Jahrbücher für Systematik, Pflanzengeschichte und Pflanzengeographie 15: 152.

Pea, A. 1931. Montbretias at Breccles Hall. Gardening Illustrated 53: 547.

Pearse, R. O. 1978. Mountain Splendour: the Wild Flowers of the Drakensberg. Howard Timmins, Cape Town.

Perrier de La Bâthie, H. 1939. Iridaceae, *in* Flore de Madagascar et des Comores. Imprimerie Officielle, Tananarive.

Peters, W. C. H. 1864. Naturwissenschaftliche Reise nach Mossambique, Band 6, Botanik, II Abteilung. Georg Reimer, Berlin.

Pfitzer, W. 1911, 1913, 1953. [Catalogues] Stuttgart.

Phillips, E. P. 1941. A note on N. E. Brown's sub-division of the genus *Antholyza* Linn. Bothalia 4: 43–44.

Planchon, J. E. 1851. *Crocosmia aurea.* Flore des Serres et des Jardins de l'Europe 7: 161, plate 702.

Plukenet, L. 1692. Phytogeographia. Published by the author, London.

Prichard, M. 1912, 1935, 1929, 1936. [Catalogues] Christchurch, England.

Redouté, P. 1813. *Antholyza prealta, in* Les Liliacées 7: plate 387.

Robinson, W. (editor). 1884. *Antholyza paniculata.* The Garden 26: 352.

Robinson, W. 1933. The English Flower Garden, ed. 15. John Murray, London (reprinted 1984, Hamlyn, London).

Roemer, J. J., and J. A. Schultes. 1817. Systema Vegetabilium, vol. 1. Cotta, Stuttgart.

Rudall, P. J. 1995. Anatomy of the Monocotyledons, vol. 8. Iridaceae. Clarendon Press, Oxford.

Rudall, P. J., and P. Goldblatt. 1991. Leaf anatomy and phylogeny of Ixioideae (Iridaceae). Botanical Journal of the Linnean Society 106: 329–345.

Salisbury, R. A. 1812. On the cultivation of rare plants, etc. Transactions of the Horticultural Society, London 1: 261–366.

Scott Elliot, G. 1890. Ornithophilous flowers in South Africa. Annals of Botany 4: 265–280.

Simpson, W. H. 1937. Plant Specialities Catalogue, page 10. W. H. Simpson and Sons, Birmingham, England.

Smith, T. 1914. Montbretias. The Gardeners' Magazine 57: 599.

Smith, T. 1915. New montbretias as pot plants. The Gardeners' Magazine 58: 390.

Stevens, H. 1936. *Antholyza paniculata.* Gardening Illustrated 58: 608.

Tenore, M. 1845. Catalogo delle Piante Che Si Coltivano nel R. Orto Botanico di Napoli. V. Puzziello, Naples.

Thomas, G. S. 1982. Perennial Garden Plants, ed. 2. J. M. Dent and Sons, London.

Thomas, G. S. 1987. The Complete Flower Paintings and Drawings of Graham Stuart Thomas. Thames and Hudson, London.

van Dijk, J. M. 1995. [Catalogue] Hillegom, the Netherlands.

van Scheepen, J. (editor) 1963. International Checklist for Hyacinths and Miscellaneous Bulbs. KAVB, Haarlem, the Netherlands.

van Scheepen, J. 1991. International Checklist for Hyacinths and Miscellaneous Bulbs. KAVB, Hillegom, the Netherlands.

van Tubergen. C. G. 1890–1930. [Catalogues.] Haarlem, the Netherlands.

Veldkamp, J. F. 1997. Overlooked genera and species in the Malesian flora: the case of *Crocosmia* (Iridaceae) and some others. Flora Malesiana Bulletin 11(8): 511–514.

Vogel, S. 1954. Blütenbiologische Typen als Elemente der Sippengliederung. Botanische Studien 1: 1–338.

Wallace, R. W. 1905, 1906, 1913, 1926, 1932. [Catalogues] Wallace and Company, Colchester, England.

Watson, W. 1893a. *Tritonia aurea*. The Garden 44: 462.

Watson, W. 1893b. *Crocosmia aurea*. Garden and Forest 6: 534.

Wijnands, D. O. 1986. The correct citation of *Montbretia crocosmiiflora*. Bothalia 16: 51.

Wolley-Dod, C. 1882. *Crocosm[i]a*. The Gardeners' Chronicle 12: 500.

Wolley-Dod, C. 1901. Hybrid montbretias. The Gardeners' Chronicle 30(2): 338–339.

'Yorkshireman, a'. 1906. Westwick Hall and Westwick montbretias. The Gardeners' Chronicle 40: 222–223.

Index

About the Authors

Peter Goldblatt is Curator of African Botany at the Missouri Botanical Garden, St. Louis. He has been associated with that institution since 1972, when he left a teaching position at the University of Cape Town, South Africa. He has made the study of the *Iris* family a lifetime research interest and is one of the world's leading experts on the family and its close relatives. His research on various genera of Iridaceae has taken him on numerous field trips to Africa as well as southern Europe, the Middle East and various parts of North America. His field work has been funded by the U.S. National Science Foundation and the National Geographic Society. Author of some 200 scientific papers, he has also written several botanical monographs, including *The Moraeas of Southern Africa* (1986), *The Genus Watsonia* (1989), *The Woody Iridaceae* (Timber Press, 1993), *Gladiolus in Tropical Africa* (Timber Press, 1996), and *Gladiolus in Southern Africa* (Fernwood Press and Timber Press, 1998), the last in collaboration with John Manning. In 1999 Peter was awarded the International Bulb Society's Herbert Medal for his contribution to the understanding of the taxonomy and biology of the *Iris* family. His most recent book, *The Color Encyclopedia of Cape Bulbs,* coauthored with John Manning and Dee Snijman, was published in 2002 by Timber Press.

John Manning was born in Pietermaritzburg, South Africa, and has been a research scientist in the Compton Herbarium at the National Botanical Institute, South Africa, since 1989. He works at Kirstenbosch National Botanical Garden in Cape Town, one of the world's great botanical gardens and an important center for research on the African flora. Although he has studied the anatomy, embryology and seed development of plants in diverse families, including the Fabaceae, Proteaceae and Stilbaceae, he has focused his research more recently on the Iridaceae, collaborating on various research projects with Peter Goldblatt. Together they have investigated the evolution and pollination biology of the African genus *Lapeirousia* and the systematics, pollination systems and evolution of *Gladiolus* in southern Africa. John and Peter have coauthored several books, including *Gladiolus in Southern Africa* and various wildflower guides to the southern African flora, the most recent of which was *Wildflowers of the Fairest Cape* (Redroof Design and Timber Press, 2000). John is also an accomplished botanical artist and photographer; his drawings have been published in numerous books and scientific journals. His most recent book, *The Color Encyclopedia of Cape Bulbs,* was coauthored with Peter Goldblatt and Dee Snijman.

GARY DUNLOP is an architect with an interest in a wide range of plants, especially Asiatic, Southern Hemispheric and woodland plants. He has put together significant collections of many unusual and neglected as well as common genera, ranging from South African sun lovers through antipodean alpines to moisture- and shade-loving plants mostly from the Northern Hemisphere. He now gardens on an exposed hilltop that has relatively shallow, lightly acid soil with intrusions of the underlying dolerite forming natural rock features. The garden is located in County Down, Northern Ireland, just above Newtownards at the northern end of Strangford Lough, where a surprising diversity of plants thrives despite, at times, the climate. His time is also spent researching many of the genera he has collected, particularly those that are not well covered in horticultural or even botanical literature. His publications include a number of articles on a diversity of plants.

One of the world's finest botanical artists, **AURIOL BATTEN** was born in Pietermaritzburg, South Africa. Always interested in plants as well as in drawing and painting, she studied botany, geography and art at the University of Natal. She became a teacher at schools in Natal, and while working in Durban she came under the influence of Niles Andersen, a well-known artist there. She extended her art training at Durban Technical College. Moving to East London in Eastern Cape province in the late 1940s, she joined forces with Hertha Bokelmann in a project to paint and document the wildflowers there. They published *Wild Flowers of the Eastern Cape Province* in 1966, and in 1967, joined by M. Courtney-Latimer and G. G. Smith, *The Flowering Plants of the Tsitsikama Forest and Coastal National Park*. Her widely acclaimed volume, *Flowers of Southern Africa* (1986), established her as a major botanical artist; she was awarded a Gold Medal by the Royal Horticultural Society. Her illustrations in *Dierama, the Hairbells of Africa* (O. M. Hilliard and B. L. Burtt), published in 1991, again showed her skill with watercolor; her habitat sketches in pencil in that volume as well as in *Flowers of Southern Africa* and the present volume give her work remarkable distinction. Her most recent paintings of *Gladiolus* were published in *Gladiolus in Southern Africa*. She continues to paint for the series *Flowering Plants of Africa,* and her work has been exhibited in London and St. Louis as well as South Africa.